AQA Environmental Studies

AS **A2**

Exclusively endorsed by AQA

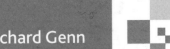

Richard Genn

Nelson Thornes

Published in 2009 by:
Nelson Thornes Ltd
Delta Place
27 Bath Road
CHELTENHAM
GL53 7TH
United Kingdom

09 10 11 12 13 / 10 9 8 7 6 5 4 3 2 1

A catalogue record for this book is available from the British Library

ISBN 978 1 4085 1390 3

Cover photograph: Photolibrary/ BrandX Pictures
Page make-up by Hart McLeod Ltd, Cambridge

Printed and bound in Croatia by Zrinski

Acknowledgements

I wish to acknowledge the unending support, guidance, encouragement and tolerance of my wife, Angela. I dedicate this book to my sons David and Daniel in the hope that their world can recover from the damage caused by the ignorance and poor choices of my generation.

I would also like to recognise the inspiration of past colleagues: Stuart Miles and Andy Kane; and the unstinting support of present colleagues: Sue Cavey, Jane Banks, Steve Jenkins and Alison Marchant.

I also wish to thank the following for permission to use their photographs in this book: Vladimir Rodriguez, Chris Hatten, Victoria Wilson, Sue and Alan Bickley, and Angela, Daniel and Gordon Genn.

The authors and publisher are grateful to the following for permission to reproduce the following copyright material:

Page 25: The stated aims of Natural England are from Natural England www.naturalengland.org.uk © Crown Copyright reprinted under Crown Copyright PSI License C2008000256; Page 30: The activities of the RSPB from The Royal Society for the Protection of Birds www.rspb.org.uk; Page 45: Case study, the Rio Bravo rainforest reserve from www.pfbelize.org; Page 286: Table 16.1, The ecological footprints and biocapacities of some areas of the world, from http://assets.panda.org/downloads/living_planet_report_2008.pdf published in October 2008 by WWF–World Wide Fund For Nature (formerly World Wildlife Fund), Gland, Switzerland; Pages 287–8: Attempts to achieve sustainable development: adapted from United Nations Conference on the Human Environment 1972 (the Stockholm Conference); The World Commission on Environment and Development 1983 (the Brundtland Commission); The United Nations Conference on Environment and Development 1992 (the 'Rio Summit'); The World Summit on Sustainable Development 2002 (the Johannesburg Summit or Rio + 10); Page 293: Table 16.4, Fossil fuel reserves – date selected from World Coal Institute data http://www.worldcoal.org/pages/content/index.asp?PageID=100 and and BP Statistical Review of World Energy 2008 www.bp.com/productlanding.do?categoryId=6929&contentId=7044622; Page 303: Case study, Management of the Exe Estuary; see www.exe-estuary.org

Every effort has been made to contact the copyright holders and we apologise if any have been overlooked. Should copyright have been unwittingly infringed in this book, the owners should contact the publishers, who will make corrections at reprint.

The continued presence of oxygen in the atmosphere relies on processes that continually replace it as other processes remove it. A small amount of oxygen was released by the **photolysis** of water. Photosynthetic bacteria, then, much later, photosynthetic plants gradually released much larger amounts of oxygen.

This oxygen absorbed UV light from the sun, causing the oxygen molecules to split. This produced monatomic oxygen that reacted with diatomic oxygen to produce ozone. These reactions allowed the ozone layer to develop, which provided protection from UV light to living organisms on Earth. Before the ozone layer developed, abundant life was not possible on land so early organisms lived in the oceans where water protected them from the UV light.

Atmospheric carbon dioxide

Carbon dioxide is naturally released into the atmosphere by volcanoes. It is an essential **greenhouse gas**, helping to retain heat in the atmosphere. Without it, the Earth would be too cold to support life. However, excessive carbon dioxide levels would cause temperatures to rise too high for life to survive.

The light output of the sun increases by about 10 per cent every billion years, so it is about 30 per cent brighter now than when life first developed. If the carbon dioxide content had remained constant, then the Earth would have heated up and life would have died out, but living organisms have helped to maintain suitable atmospheric temperatures by removing carbon dioxide from the atmosphere and storing the carbon in fossil fuels and carbonate rocks such as limestone and chalk.

The water cycle

Heat energy from absorbed sunlight causes water in the sea to evaporate. This may be carried over land where it falls as rain then flows back to the sea. Most rain falls relatively close to the coast, but **transpiration** by plants returns water vapour to the atmosphere so it can be blown further inland.

Transpiration from leaves is unavoidable as the moist stomata must be open to allow exchange of gases such as carbon dioxide. It is important in cooling plants and in the upward transport of dissolved minerals.

The individual features of the Earth and their inter-connections can be very complex. Although they are interdependent, it is often easiest to focus on one aspect at a time, such as the **biosphere**, **hydrosphere**, atmosphere or **lithosphere**.

Activity

This is an imaginative activity that will help you to concentrate on important issues.

A traveller from another galaxy arrives on Earth – having visited many planets with no life – and asks you: 'What is so unusual about your planet that it has so much life and such a variety of life forms?'

List what you think are the five most important features of Earth and add a brief explanation.

Did you know?

The anaerobic bacteria that lived before oxygen was abundant were almost completely wiped out by the oxygen that is toxic to them. They still survive in anaerobic environments such as swamps, rice *padi* fields, landfill sites and mammal intestines where they release methane gas.

AQA Examiner's tip

Many processes on Earth produce dynamic equilibria where a range of processes cancel each other out to produce stability. Look for them in the water, nitrogen, carbon and phosphorus cycles, in population regulation and many other areas of the subject.

Summary questions

1. Outline how the following features of the Earth have provided conditions that allowed life to develop:
   - position in the solar system
   - structure
   - rotation. *(2 marks each)*

2. Outline three ways in which the properties of water allow living organisms to survive on Earth. *(2 marks each)*

3. Explain why the temperature range found on Earth is suitable for the survival of living organisms. *(2 marks)*

4. Explain how the presence of life has controlled the conditions on Earth and made it possible for more habitats to be colonised. *(6 marks)*

2 Wildlife conservation

Learning objectives:

- ■ Why is wildlife conservation important?
- ■ How do we threaten it?
- ■ What can we do to conserve wildlife?

The rationale for wildlife conservation

There are many reasons for conserving wildlife, some being based on subjective opinion, while others are based on more objective facts that can be used in scientific debate.

Knowledge and understanding

By studying wildlife, we can learn a great deal of interesting information that is useful in understanding ourselves, why wildlife conservation is important and how successful conservation strategies can be planned.

Human behaviour

By studying other primates, particularly the great apes such as gorillas and chimpanzees, we can learn about social structures and group behaviour.

The interdependence of life on Earth

If we understand how species affect each other then we may also understand how our survival relies upon the survival of the other species with which we share the Earth.

Biomimetics

Biomimetics is the knowledge of how other species are adapted to survive and the application of that knowledge to solving human engineering problems. In the table below are some examples of the practical application of biomimetics.

Key terms

Biomimetics: the study of living organisms so the knowledge gained can be applied to engineering or technological developments.

Fig 2.1a and b *The design of aircraft wingtip fins (a) was based on bird wings (b)*

Table 2.1 *Examples of the use of biomimetics*

Species adaptation	Application in human engineering
The bones of birds must have thin walls to make them light. Their strength is retained by having internal cross-trusses to prevent the bone bending and snapping	Similar trusses are used inside tubular bridges so that they can be strong but light
Humpback whales swim in tight circles to surround shoals of food fish. This is made easier by the shape of their pectoral fins, which have scalloped leading edges	This feature is being applied to rudder design to improve ship manoeuvrability
Shark skin has grooves at right angles to the direction of water flow. This reduces friction and makes swimming easier	Surface grooves called riblets are being developed for aircraft, swimsuits and boats. Friction can be reduced by up to 15 per cent, which results in lower fuel consumption
Soaring birds, such as eagles, storks and vultures, use spread wingtip feathers to reduce the flow of air from underneath the wing where the air pressure is higher, to above the wing. This reduces air vortices and increases the lift of the wing	Aircraft now have fins fitted to their wings to achieve the same effect

Aesthetics and recreation

A world that is rich in wild animals and plants is a more pleasant place to live, but our enthusiasm for conservation is not the same for all species.

Furry, cuddly and appealing species such as pandas, monkeys and penguins are more popular than spiders, snakes and slugs, for example. However, species do not live in isolation from each other. The more aesthetically pleasing species may require the services of less popular ones to survive. Most people enjoy seeing wildlife as it increases their quality of life. This often creates economic benefits at visitor attractions such as zoos and wildlife parks, through membership of conservation organisations or through ecotourism activities such as bird-watching and whale-watching.

Morals

Many people believe other organisms have a right to life and that it is wrong to kill unnecessarily. Public concern is greatest when high-profile animals such as whales, tigers or elephants are threatened. Most people would support the right of other species to exist, but fewer would argue that malaria mosquitoes, parasitic worms or rats have a similar right. So, the moral argument for wildlife conservation is based on subjective opinion rather than objective fact.

Ethics

In more affluent societies it is not necessary to exploit wildlife for food, but in some societies it is an essential part of the diet, such as the collection of wild birds' eggs and 'bushmeat'. In less economically developed countries many people may have no alternative source of food and it would be difficult to convince someone that they have less of a right to live than their source of food has.

Economic reasons for conservation

We can easily forget the benefits we have gained in the past from wild species and that also all domesticated or cultivated species were once wild. We may also be ignorant of the enormous potential we are currently losing, as species that we have never studied become rarer and eventually extinct.

Medical benefits

Physiological research

Many species have been used for physiological research in such diverse areas as nerve function, leprosy and drug **teratology** (the study of non-inherited birth abnormalities).

Did you know?

'Flagship species' are the most appealing or notable species that can be used to highlight conservation campaigns or vulnerable habitats. If the flagship species are successfully protected then the other species that share its environment may also be protected.

 Examiner's tip

The cost of wildlife conservation programmes is often a major obstacle to successful conservation. However, the economic potential of exploiting wildlife means we may lose more money by not conserving it.

Key terms

Teratology: the study of the causes of birth defects.

Table 2.2 *Some species used in physiological research*

Area of physiological research	Species used
Nerve function	Squid have wide diameter nerve cells that are easier to study than mammal nerves. Research on them has helped improve our understanding of cell membrane ion pumps and therefore problems including heart disease, stroke and Alzheimer's disease
Leprosy	Armadillo are among the few animals that can catch leprosy and are used in the study of the disease and the production of vaccines to prevent it
Drug teratology	Sea urchin embryos can be used to investigate whether drugs are likely to cause birth abnormalities. The purple sea urchin of North America is common and a single female can produce half a million eggs. Drugs that would cause birth abnormalities in humans also cause abnormal sea urchin embryo development

Fig 2.2 *Yew trees produce taxol, which is extracted for use in the treatment of breast cancer*

■ Key terms

Indigenous species: species that are native to the area.

Development of new medicines

Many plants, especially those in the tropics, produce biologically active chemicals such as alkaloids that may kill the pests that eat them. In controlled doses they may have medical uses in treating human diseases.

Table 2.3 *Examples of medicines extracted from other species*

Species	Medicine
Poppies	Painkillers morphine and codeine
Mexican wild yam	Diosgenin, derivatives of which have been used to make many steroid medicines including the first contraceptive pill
Cinchona tree from South America	Quinine for malaria protection
Rosy periwinkle from Madagascar	Vincristine and vinblastine, used in cancer treatment
Yew tree	Taxol, extracted to treat breast cancer

A very small proportion of the plants that exist have been studied for the medicinal substances they may contain. This is a powerful argument for conserving all the other species or, preferably, entire habitats.

Food resources

Very few species are currently exploited for food. About 3,000 plant species have ever been used for food, which is about 1 per cent of the total number of known plant species, and only about 150 of these are important. Only a few dozen animal species are important for food.

People often destroy or damage natural ecosystems and remove whole communities of wild species to allow the introduction of a very small number of domesticated crops and livestock.

New food species

Indigenous species are usually better adapted to the local climate, pests and soil conditions than introduced species and therefore may give better yields. Selective breeding may be needed to enhance desirable characteristics and eliminate undesirable ones so species that have no obvious use could become very valuable.

Most of the species that are currently farmed were domesticated a long time ago, but there have been some recent attempts to domesticate new species, including such animals as bison, eland, ostrich, cane rat and giant snail.

In Papua New Guinea there are 250 plant species that yield edible fruit, but fewer than 50 are currently cultivated.

There are 1,500 plant species in the spinach family. They grow well in salt-rich soils and could be grown in areas where irrigation has caused soil salinisation.

A perennial variety of maize was found in Mexico (in an area where the vegetation was due to be cleared). Because it is perennial it grows year after year, which removes the need to buy new seeds annually. The reduced need for ploughing and cultivation also reduces the risk of soil erosion and land degradation.

Wild varieties for breeding programmes – protecting the gene pool

For commercial crop species, relatively few varieties of crops are grown and each variety may be quite uniform genetically. If it is necessary or desirable to breed new characteristics into the crop then it will probably be necessary to look outside the highly bred commercial varieties. Varieties grown in subsistence farming areas or wild varieties are more likely to hold these desirable characteristics that are not found in the commercial varieties.

Gene pool problems

The **gene pool** is the total number of different genes in all of the individuals in a population. A small population with great genetic variety may have a larger gene pool than a large population of genetically similar individuals.

Domesticated species are often inbred, having been produced from a very small number of original ancestors. They lack the wide variety of characteristics found in wild or commercially unimportant cultivars (varieties). They may therefore be less able to cope with changes in conditions beyond the limited range to which they are all adapted. There is also a greater risk of disadvantageous recessive genes causing problems.

The need for a large gene pool means that it is not enough to protect a few representatives of each species. A wide range of genetically different, relatively distantly related individuals must be protected.

Each region within the geographical range of a species will have its own gene pool with some genes that are unique to that area. These will exist because of the need to be adapted to the specific conditions in that region. So, to protect as much of the gene pool as possible, each species should be protected over its entire range, not just in a few convenient areas.

The 19th-century Russian zoologist Nikolai Vavilov identified certain areas of the world with especially large numbers of plant species of economic importance and where these populations had great genetic diversity. These areas are called **Vavilov Centres**.

Many desirable characteristics are known to exist in wild gene pools:

- disease-resistant rice and potatoes
- drought-resistant maize
- cold-resistant pineapples
- species for biological control.

Unfortunately, many Vavilov Centres are in areas that are suffering environmental destruction and degradation.

Species used for pest control

Many wildlife species can be used to control agricultural pests. Some examples are parasitic wasps and predatory beetles.

Other material resources

Humans obtain a wide range of other materials from animals and plants:

- wood for construction, paper, fuel, resins, tannins
- insecticides – pyrethroids, rotenone
- fibres – cotton, wool, flax, hemp, sisal
- cosmetics
- dyes
- skins
- oils.

> ### Key terms
>
> **Gene pool:** the total variety of different genes in all the members of a population.
>
> **Vavilov Centre:** an area of the world, identified by the Russian zoologist Nikolai Vavilov, where crop plants were first domesticated and where wild varieties are still found.

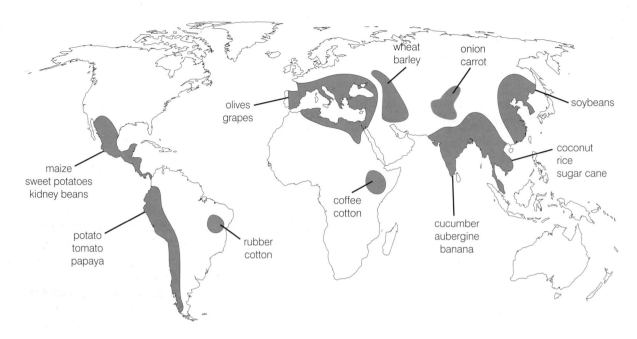

Fig. 2.3 *Important crop species in Vavilov Centres*

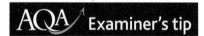
The Earth's life-support systems

Planet Earth is our only home. The life-support systems that allow life to survive on Earth are partly maintained by living organisms. As human numbers and our individual impacts on the planet increase, it becomes more important to understand how the life-support systems of Earth function and how we may affect them.

Maintenance of the atmosphere

The organisms that photosynthesise (photoautotrophs) remove carbon dioxide from the atmosphere and release oxygen. This is a vital service for all forms of life on Earth, not just those that are aerobic and require oxygen for respiration. If carbon dioxide levels had not been reduced from their original level then the greenhouse effect would have made the Earth too hot for life to exist.

Oxygen in the stratosphere produces ozone that absorbs ultraviolet light.

The role of plants in the hydrological cycle

Plants increase humidity by returning water to the atmosphere through transpiration and by the evaporation of rainwater that landed on leaves (interception). This rise in humidity increases future rainfall downwind.

Soil formation and conservation

Dead organic matter from plants and animals provides nutrients for future plant growth but the nutrients are only released by the actions of other organisms. Detritivores such as beetles, millipedes and woodlice break up the dead organic matter, which is then broken down by the enzymes released by decomposers such as bacteria and fungi.

The foliage and roots of the plants help to protect soil by reducing the erosion caused by wind and water.

Species interdependence

The survival of many species relies upon the services that are provided by other species.

Food supplies

Food species provide an obvious service to the species that eat them. Predators may also benefit the species they eat by removing the sick and weak.

Pollination

The flowers of many plants are pollinated by animals including insects, birds and bats. This is a much more reliable method than wind pollination as the pollinating animal will deliberately seek out the flowers for their nectar and therefore transfer the pollen between flowers.

Many flowers have evolved to attract particular insects. Equally, many insects have evolved to feed from particular plants.

Seed dispersal

Wind dispersal of seeds is haphazard and does not usually carry the seeds very far. The seeds have to be small to travel very far, which reduces the amount of food energy carried in the seed and therefore its ability to produce a seedling plant.

Animal dispersal is more reliable. Fruits often have brightly coloured, tasty flesh to attract animals when the seeds are ready to be spread. The seeds may pass through the animal's intestine and be delivered to a suitable habitat in faecal material, which acts as a fertiliser. The seeds of many species are stimulated to germinate by their passage through the gut.

Habitat provision

One species may provide a habitat for others, such as birds nesting in trees or small animals sheltering under logs or in leaf litter.

Activity

You meet someone that isn't interested in wildlife and doesn't care that species are becoming extinct.

What would you say to them to try to convince them that wildlife conservation is important for everyone, including them?

AQA Examiner's tip

Make sure you can give a well-balanced explanation of why wildlife conservation is important, with the full breadth of reasons backed up with examples for each.

Summary questions

1 Explain why wildlife conservation is important for each of the following reasons:
- aesthetics
- medical research
- genetic resources for agriculture
- biological pest control in agriculture
- nutrient cycling. *(3 marks each)*

2 Suggest ways in which:
- animals depend on green plants
- green plants depend on animals
- all living organisms depend on decomposers. *(2 marks each)*

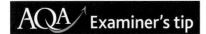

AQA Examiner's tip

When you choose to write about examples of species threatened by human activities, make sure you use a range of threats from each of the categories. Don't just choose the high-profile species that are deliberately exploited, such as whales, tigers and rhinos.

■ How humans threaten wildlife

Humans threaten wildlife in three main ways:

- unsustainable exploitation for human gain – the species are not necessarily threatened just by being exploited, but if they are removed in such large numbers that the population cannot sustain itself then the species will become rarer
- eradication because they threaten or interfere with human activities
- particular species cannot survive the environmental changes caused by human activities.

Unsustainable exploitation for human gain

Food

A wide range of species have become rarer or even extinct because they have been hunted and overexploited for food. These include the dodo, passenger pigeon, cod, swordfish, tuna and many sharks.

Fashion

Many species have been collected or hunted because people find their skins, shells or horns attractive.

Table 2.4 *Examples of species hunted for fashion*

Animal	Product
Crocodiles and alligators	Leather for bags and shoes
Turtles	Shells for ornaments and the frames for glasses (tortoiseshell)
Rhino	Horn for dagger handles
Shellfish	Seashells for tourist souvenirs
Elephants	Ivory
Wild cats (e.g. jaguar, ocelot, snow leopard)	Fur coats

■ Did you know?

Banning exploitation may protect the species but remove a source of income. Some species, such as alligators and crocodiles, can be farmed for their skin which provides an income without threatening the wild population.

■ Case study

The snow leopard

Snow leopards are threatened by habitat loss and loss of prey species, but the main threat is hunting for their skins. A snow leopard coat would require six to 10 skins and could be worth £30,000 on the black market.

Fig. 2.4 *The snow leopard is still hunted illegally for its fur*

Pets and entertainment

Many species have been collected as pets or for entertainment purposes. Species that are popular as pets include:

- tortoises
- tropical fish
- parrots
- lizards
- snakes.

Case study

The yellow-headed parrot

Yellow-headed parrots are social, easy to tame and mimic human voices well, which makes them very popular in the pet trade. They are protected by **CITES** Appendix I, but are still poached illegally by cutting into the nesting trees to take eggs and chicks. The population has dropped by 95 per cent since the 1970s.

Fig. 2.5 *A wild yellow-headed parrot (left) and a nesting tree that poachers have robbed*

Sea animals such as dolphins and orcas (killer whales) are collected for public entertainment in dolphinaria. Some plants are collected as house plants, including 'air plants', some tropical exotics and many insectivorous plants.

Furniture and ornaments

Many species have been selectively collected for making furniture or for ornamental purposes.

Table 2.5 *Examples of species collected for furniture and ornaments*

Species	Product
Mahogany and teak	Furniture
Ivory	White piano keys
Ebony wood	Black piano keys
Coral	Jewellery
Shells from turtles and tortoises	Jewellery

Traditional medicines

The belief in traditional medicines in some cultures has created a high demand for products of some species that are becoming rare as a result.

Table 2.6 *Some species used for traditional medicines*

Species	Traditional remedy
Tigers	Different parts of the body are used as cures for different problems: claws as a sedative; the tail for skin diseases; dung for alcoholism; brain for laziness; whiskers for toothache. There is no conclusive scientific evidence that they work
Snakes	Traditional medicines are made from snakes in the belief they will cure arthritis and skin disease
Bear bile	Bears are caught and kept in cages so the bile can be drained from their gall bladders. It is used for sufferers of fever and liver disease

AQA Examiner's tip

Make sure you understand the difference between the conservation of wildlife to research genuine medicines and the unsustainable exploitation of animals to produce high-value products of mythical value but no proven effect.

Fig. 2.6 *All rhinoceros species are threatened by poaching*

Case study

Rhinoceros horn

Rhinoceros horn has been used in traditional medicines for hundreds of years as a supposed cure for many medical problems ranging from nosebleeds to fevers and smallpox. They are also used to make ornamental dagger handles. All species are protected by CITES Appendix I, but they are still hunted illegally.

Other products

A very fine oil has been extracted from whale blubber and from the spermaceti in the heads of sperm whales. It was used until the 1970s in the manufacture of candles, soap, lubricating oil and cosmetics. It has now been replaced with other oils including oil from the jojoba plant.

Eradication of predators and competitors

Many species have been persecuted because they threaten us or conflict with human activities:

- species that threaten humans – sharks, poisonous snakes
- predators that threaten livestock – wolves, lions, birds of prey, herons
- agricultural pests – insects, fungi, molluscs
- disease risks – malaria mosquitoes, tsetse flies
- pests in forestry plantations – wood-boring insects, deer, beavers
- wild herbivores that eat crops or compete with livestock – rabbits, deer.

The inability to survive habitat alteration

Species that are threatened by deliberate exploitation or eradication are often very obvious, but most endangered species are threatened by changes to their habitats.

Unintentional deaths caused by human activities

The deaths of species caused by human activities include dolphins killed as a by-catch of tuna fishing or albatrosses caught on tuna long-lines, and animals killed in fields during crop harvesting and as roadkill.

Case study

The barn owl

Barn owls hunt for small mammals in grassland, including roadside verges. Because the owls are so light they can be pulled in behind one passing vehicle and be hit by the next. Some councils cut verges very short where barn owls live so they will not hunt next to the roads.

Fig. 2.7 *A barn owl killed by passing traffic*

Introduced species

The community of species found in an area will be adapted to their abiotic and biotic surroundings. An introduced species may have a competitive advantage, which means that the indigenous species cannot survive.

Many of these introductions have had a catastrophic effect on the indigenous species. This is especially true for isolated areas such as islands where there is an odd community of species that managed to colonise from other land masses a long way away and have evolved from a few original species to occupy a wide variety of available **niches**. They may have evolved in the absence of mammal predators and are not adapted to cope with them if they are introduced.

Predators

Some introduced species have become predators of the indigenous species. Some examples of indigenous species threatened by introduced predators include the following.

- In UK rivers water voles are killed by American mink that escaped from fur farms.
- Many species in Australia are threatened by the cane toads that were introduced from South America to control insect pests in sugar cane plantations. They are general predators, eating lizards, insects, small marsupials and ground-nesting birds.
- The small Asian mongoose was introduced to Hawaii to control rats that were pests in sugar cane fields. They have devastated **endemic** ground-nesting birds.
- The ground-nesting birds on many oceanic islands such as New Zealand, Australia and Hawaii are threatened by introduced cats, rats, pigs and dogs.
- The vegetation of many islands, such as Round Island in the Indian Ocean, has been seriously damaged by introduced herbivores such as rabbits and goats. They were introduced deliberately by sailors hundreds of years ago so they would be 'living larders' for future visits.

Case study

The Nile perch

The Nile perch was introduced into Lake Victoria in Africa in the 1950s to improve food supplies, but it ate the indigenous fish species such as cichlids, some of which are now extinct. Overfishing has now reduced Nile perch numbers. Local people used to dry cichlid meat using the heat of the sun. The Nile perch is much larger and the meat is preserved by smoking, which requires wood. The extra demand for wood has increased deforestation and soil erosion.

Key terms

Niche: the niche of a species is the role it plays in its habitat, which includes how it makes use of resources and responds to the other species in its habitat.

Endemic: an endemic species is indigenous to a particular area and is not naturally found elsewhere. (The word is also used in disease epidemiology to mean a disease that is normally present.)

Fig. 2.8 *An endemic cichlid from central Africa*

■ Activity

Make a spider diagram of the different ways that introduced species threaten indigenous species.

Around each threat give examples, with details of named species that have been introduced or are threatened.

Competitors

Some introduced species occupy the same niche as indigenous species and may out-compete them. Examples of introduced competitors that threaten indigenous species include the following.

- ■ The grey squirrel was introduced into the UK from North America in the 19th century. It out-competes the indigenous red squirrel as it is better at using the available food. It is better at digesting acorns from oak trees. Being larger it can compete successfully for nest sites. It also carries the squirrel pox virus that kills red squirrels.

- ■ Rhododendrons were introduced to the UK from Asia for their ornamental flowers and as cover for game birds. They are very invasive and, being evergreen, they shade the ground and reduce the growth of native vegetation.

- ■ The harlequin ladybird was introduced to Europe from Asia, reaching Britain in 2004. It is a voracious predator and out-competes indigenous ladybird species. They also predate other ladybird species.

Fig. 2.9 *Ship ballast water can introduce species such as seaweeds and crustaceans into new areas*

Disease

Indigenous species may have no resistance to introduced diseases. Examples include the following.

- ■ Elm trees in Europe and North America have been killed by a fungus that was accidentally introduced from Asia and has been spread by a bark beetle.

- ■ More than half the indigenous bird species of Hawaii are endangered or have become extinct in the last 100 years as a result of habitat loss and the impact of introduced species. Pigs, cats, snakes, rats and mongooses are predators of ground-nesting birds. Hawaiian honeycreepers are threatened by the malaria that is carried by introduced mosquitoes.

- ■ The signal crayfish was introduced to the UK from North America to provide food. It carries a fungus disease that kill the indigenous white-clawed crayfish.

Changes in abiotic factors

When human activities change the environment it may become unsuitable for the species that live there.

The species that are most likely to be affected are those with specialised requirements, which are often the species that were originally rare before

Fig. 2.10 *An endangered Iiwi honeycreeper from Hawaii*

Fig. 2.11 *Sundews are only found in waterlogged soils*

Table 2.7 *How changes in abiotic factors threaten species*

Threatened species	Changes in abiotic factors
Fish such as trout must lay their eggs in gravel in well-oxygenated parts of rivers where the water is flowing fastest	Rivers that are straightened to reduce the risk of flooding may become unsuitable for trout. The uniform depth and flow rate produce a silty riverbed without the faster-flowing stretches that would have washed away the silt and exposed the gravel
In some tropical areas during the dry season exposed river sandbanks are important nest sites for freshwater turtles and lizards such as the green iguana	Hydro electric schemes may cause fluctuations in water levels that flood the nests and kill the eggs
Crayfish and sensitive fish species	Acid rain increases the acidity of rivers and lakes. The cell surface proteins of fish eggs and gills of sensitive fish species such as trout may be denatured so they die. Animals such as crayfish may die out as the increased acidity dissolves their exoskeletons
Species of river plants	Human land uses such as farming or mining may increase the turbidity of a river. This may reduce light penetration and prevent plants from photosynthesising
Marshland plants such as the sundew may be present because they can survive waterlogged, anaerobic soil conditions. They trap and digest insects to provide the nitrogen that cannot be obtained from the nutrient-deficient soil	Drainage schemes that produce more aerobic soil may allow more vigorous competitors to colonise the area

the habitat change threatened them with extinction. Although the habitat has not been destroyed, the conditions are no longer what they require.

The changed conditions may not kill all the individuals of a species. Howevere, if the death rate increases to a point where it is greater than the reproduction rate then the population will decline and may eventually die out.

Habitat destruction

Major changes to a habitat may cause the entire community of species to die out. Examples of activities that destroy habitats include:

- deforestation either to create farmland or to exploit the timber
- flooding caused by reservoir construction
- mining
- urbanisation.

▪ Conservation methods

There are many different ways in which humans can help to conserve wildlife but they all do one of two things: they increase the success of breeding or reduce the death rate. The best approach is to remove the threat itself but this is rarely under the direct control of those who wish to protect the wildlife. The threat may be reduced by a variety of legislative, administrative and management techniques. It is common to use a number of methods at the same time, but it is easier to study them individually.

Legal protection

Laws to protect endangered species are valuable but they are rarely the complete solution. Enforcing the law can be difficult and if the law is not

Hint

Use the other chapters to find details of how these activities alter the environment and affect wildlife.

Activity

Produce a poster or information sheet to show a range of different human activities that indirectly threaten named (and illustrated) species.

Include details of why each species has become rare. Avoid vague statements like 'they can't stand it'. Talk about changes in specific abiotic and biotic factors and how they change survival rates.

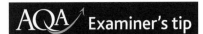
advertised or understood then it may be broken through ignorance. A law that is not enforced may be worse than no law at all as people may assume that, since the species are now legally protected, the problem has been solved.

Trade controls

It may be very difficult to prevent protected species from being collected, especially in remote areas, but if they are being sold to buyers a long way away they may have to pass through ports, airports or cross borders where there are already customs checks for other goods.

CITES – the Convention on International Trade in Endangered Species

CITES is an international agreement between governments that regulates the international trade in endangered animals, plants or their products.

■ Appendix I includes species threatened with extinction and places a complete ban on trade (except by special licence, for example for zoo breeding programmes). Species include tiger, blue whale, rhinoceros, chimpanzee and gorilla.

■ Appendix II includes species that may be threatened with extinction if trade is not closely controlled. Species include hippopotamus, polar bear, Venus fly trap, Honduras mahogany, great white shark, basking shark and whale shark. Limited trade is permitted.

■ Appendix III includes species where particular countries have requested assistance from other countries in protecting certain species, for example the walrus in Canada and the pygmy anteater in Guatemala.

Banning damaging activities

Wildlife and Countryside Act (1981 and 1984)

The **Wildlife and Countryside Act** is a UK law that provides protection for many wildlife species by legally protecting their habitats or by making the exploitation or harming of them illegal. Some features of the Wildlife and Countryside Act that protect wildlife are:

■ **SSSIs** may be designated to protect habitats

■ the uprooting of wild plants is generally illegal

■ harming wild birds or their nests is illegal except for some 'pest' or game species

■ many mammals are protected, for example otters, shrews, dormice

■ badgers and their setts (tunnels) are protected

■ bats may not be disturbed and woodworm treatment chemicals used in roofs where there are bats must not be toxic to bats.

Designation of legally protected areas

Designated protected areas place restrictions on the activities that may be carried out there to provide protection for the community of species that live there rather than just for particular species.

Examples of designated areas for wildlife conservation in the UK include SSSIs, **NNRs**, **SPAs**, **SACs**, **MNRs** and **Ramsar sites**.

Sustainable management of exploitation

If a species is being exploited at a sustainable level then the future of the species may not be threatened.

plants such as cowslips and monkey orchids, the chalkhill blue butterfly and birds including the stone curlew and skylark.

Threats to chalk grassland are:

- intensive farming with ploughing and use of fertilisers to substitute for fertile soil
- 'improvement' of the grassland using fertilisers, pesticides and re-seeding with more productive grass varieties
- urban expansion and road building
- abandonment of grazing when sheep farming is not profitable. This leads to secondary succession where thorn and scrub woodland replaces the grassland.

Fig. 3.6 *Scrub woodland encroaching into chalk grassland when grazing has stopped*

Hay meadows

Traditional hay meadows were not cut until the end of the summer when the mature dry grasses could then be stored as nutritious winter fodder for livestock. By the end of the summer the annual wildflowers had flowered and produced seeds, which would have safely fallen to the ground, thus ensuring wildflowers for future years. No weedkillers were used. No fertilisers were used, which would have made the grasses grow taller and shade the wildflowers.

Rare species that live in meadows include plants such as the fritillary and greater butterfly orchid and birds such as the corncrake.

Threats to hay meadows include the following:

- 'Improved' grassland – the ground is sprayed with weedkillers, then ploughed and re-seeded with fast-growing varieties of grass that provide a higher crop yield. The wildflowers are killed and the reduced plant diversity means there are fewer insects and birds that feed off them.
- A change in grassland management to silage cutting – the grass is cut several times during the summer and stored wet in silage clamps. The regular cutting kills the wildflowers before they can produce seeds and destroys the nests of ground nesting birds.

a

b

Fig. 3.7a and b *Traditional hay-cutting and drying; and modern silage harvesting*

Upland moorland

When the woodlands of upland Britain were cleared for fuel, the land was often used for sheep grazing, which produced a heather and grass community of great wildlife value.

Regular burning helps to maintain the moorland plagioclimax, which may be used for grouse shooting.

Threats to moorland include:

- conversion to 'improved' grassland where high yielding grass varieties are planted
- afforestation with conifers
- reservoir construction
- abandonment of grazing or grouse shooting, allowing secondary succession
- increased visitor pressure as a result of the Countryside and Rights of Way Act (2000).

Important moorland species include the hen harrier, golden eagle and black grouse.

Coasts and estuaries

Coastlines next to open oceans often have relatively stable climates because of the high heat capacity of the seawater, which resists temperature changes, especially if the coast has a consistent ocean current.

Britain has a temperate maritime climate, which is warmer than many areas at the same latitude because the Gulf Stream (North Atlantic Drift) brings warm water from the Caribbean Sea.

The coastline of the UK is over 18,000 km (11,000 miles) long and varies from mountainous areas of hard rock to lowland areas with soft sedimentary rocks. This produces a wide range of different coastal habitats. There are over 1,000 islands and many cliffs where birds can often nest on the ground with no threat from mammal predators.

Estuaries

A rise in sea level in medieval times flooded river valleys near the coast and produced a large number of estuaries of different types, depending on the shape of the valley.

Estuaries are very biologically productive because of the large amounts of dead organic matter carried downstream by rivers or brought in by the incoming tides. Shallow, warm conditions also allow rapid growth of algae.

Abiotic factors in estuaries

The abiotic features of different areas in an estuary can vary, producing locally different communities as follows:

- salinity – ranging from pure seawater to freshwater where the river flows in
- depth
- turbidity – ranging from clear water to water with fine suspended solids brought in by river or from the sea
- water flow rates – from very slow movement to speeds over 10 mph
- period of exposure to air – from almost never covered with water to almost never exposed to the air

- temperature fluctuations – especially when the tide is out as water has smaller seasonal temperature fluctuations than the air
- substrate type – from coarse gravel to fine clay.

These varying characteristics produce great habitat variety, and increase the number of species that live there and therefore their wildlife value. If just a small part of the estuary habitat is lost it might be the only part with that unique community of species.

The estuaries of Britain are particularly important as few other areas of the same latitude have such a mild climate with the topography that produces estuaries. British estuaries attract large numbers of migrant birds which come to the ice-free mud-flats, marine grasses and salt marsh feeding grounds during the winter. Birds migrate from as far away as Siberia, Greenland and northern Canada. Important estuary species include many birds such as waders, ducks and geese, and many invertebrates such as molluscs, crustaceans and worms.

Threats to estuaries include:

- port developments
- pollution from rivers that drain into the estuaries
- land reclamation
- pollution from industrial developments such as oil terminals.

Summary questions

1. Outline the roles of the governmental organisations involved in wildlife conservation in the UK. *(9 marks)*

2. Outline the functions of the different designations that protect habitats and wildlife in the UK. *(10 marks)*

3. Describe how agri-environmental schemes have reversed some of the environmental damage caused by farming in the past. *(8 marks)*

4. Outline the role of non-governmental organisations in wildlife conservation in the UK. *(6 marks)*

5. Explain how human activities have produced semi-natural habitats in the UK that are important for wildlife. *(6 marks)*

6. Describe the ecological factors that control a named UK habitat. *(6 marks)*

7. Outline the human activities and land-use changes that threaten wildlife in the UK. *(6 marks)*

4 Conservation abroad

Learning objectives:

- What are the ecological conditions in and threats to tropical rainforests, coral reefs and Antarctica?

- How might conservation efforts reduce or avoid human impact on these environments?

- How might increasing population pressure require an approach of sustainable development?

Key terms

Species: a group of organisms that resemble each other more than other organisms and naturally interbreed to produce fertile offspring.

Abiotic factors: physical factors such as light, temperature and water.

Biotic factors: biological factors such as food and disease.

There are far too many habitats in other parts of the world for us to be able to study them all. However, it is worth studying a few that are very important, especially if they are different from habitats in the UK. Tropical rainforest, coral reefs and Antarctica all have natural ecosystems that are not maintained by human activities, unlike most UK habitats. Each is threatened by human activities in a variety of different ways.

Tropical rainforest

The main ecological features of tropical rainforests

Tropical rainforests do not have large seasonal changes in climate because they are close to the Equator and the conditions are ideal for biological productivity with high light levels, warm temperatures and plenty of water. There have not been significant climatic changes for millions of years.

These stable conditions throughout the year and over long time periods have given time for a huge variety of **species** to evolve, which has produced very high biodiversity. The **abiotic factors** do not usually require very specialised adaptations but **biotic factors** and species inter-relationships have acted to increase the biodiversity. Different plant species may have evolved to become better adapted to small local differences in climate, soil or water availability. Animal species may have evolved to pollinate, feed on or spread the seeds of particular plants, and other animals have adapted to feed on them. There are other inter-relationships to do with **nutrient** supply, habitat provision and **microclimates** with changed light levels, temperatures and humidity. A number of species may occupy specialised niches, each being uncommon, instead of one common species occupying a more general niche as is often found in less diverse ecosystems.

This high biodiversity makes tropical rainforests especially important, with each area being unique and therefore important. So, the loss of a small area may result in the loss of unique species or a unique part of the gene pool.

The high light levels enable dense vegetation to develop as there is enough light for photosynthesis to take place even after much of the light has passed through many leaves.

Different plant species are adapted to live in different layers in the rainforest. The tallest trees have buttress roots to support them while forest floor plants have extra pigments to make use of the lower light levels.

Fig. 4.1 *Buttress roots support tall trees on shallow soil*

5 Life processes in the biosphere

Learning objectives:

- How can an understanding of ecology help wildlife conservation decisions?

- How is an understanding of the conditions that species require to survive important in managing **habitats** to benefit them?

- How does such an understanding make it easier to predict the likely impacts of human actions?

Examiner's tip

You do not need detailed knowledge of the biological basis of adaptations, but it is important to know how abiotic and biotic factors affect survival. It is often easiest to describe these by using examples of species with particular requirements or ones that are threatened by habitat changes.

Key terms

Habitat: the place where an organism, species or population lives.

Population: all the individuals of a species living in a particular area.

Exoskeleton: a skeleton on the outside of an organism such as found in insects and crustaceans.

There are a number of processes that it is important to study for an understanding of the environment. These are covered in this chapter.

Adaptation to the environment

To be able to survive, a species must be well adapted to its physical and biological environment.

Most organisms can only survive within a relatively narrow range of conditions called the range of tolerance. Each individual has its own range of tolerance. A **population** with a large gene pool is able to survive environmental change more easily as it is more likely that some individuals will be adapted to the new conditions. Although those that were not adapted to the new conditions would die, the survivors would be able to breed and rebuild the population.

Abiotic factors

The survival and distribution of species is largely controlled by abiotic factors. Being able to survive extreme conditions to which few species are adapted may avoid competition with other species that won't be able to live there, but it may also limit the areas that will be suitable for that species and make it vulnerable to environmental change.

Temperature

No organisms can survive where low temperatures cause the cells to freeze or high temperatures denature their proteins. Within this range a species will be able to survive if their proteins, including enzymes, are functional. Even within the range of temperatures that is generally suitable for life, each enzyme will function within a particular temperature range. This is a problem for organisms that live where temperatures fluctuate a lot, but the problem can be overcome if the body's internal temperature can be kept constant. Lizards achieve this through their behaviour: sunbathing to warm up, sheltering to avoid overheating and hibernating if they cannot get warm enough. Birds and mammals generate heat through the respiration of stored food. This greatly increases their food requirements but allows them to live in more varied environments.

Bacteria around volcanic vents have evolved to survive at temperatures over $110\,^{\circ}C$.

Light

Light that is absorbed during photosynthesis is the source of energy for most food chains. Plants have evolved pigments to absorb light in particular habitats. Many woodland floor plants have red or blue pigments to make use of the green light that passes through the canopy vegetation. These plants often cannot survive continual bright sunlight.

UV light damages living cells and is particularly dangerous to animals with thin skins, such as amphibians.

Did you know?

■ In a nature reserve where there are lizards it may be necessary to provide small, sunny clearings where they can sunbathe to warm up during cool weather. Sunny, sandy areas are ideal for egg laying as the eggs will be warm and are able to develop.

■ Lady's slipper orchids need intermittent sunlight so they survive best in woodlands with a fairly open canopy. Habitat management may involve selective felling, pollarding or coppicing.

Fig. 5.1 *Many frog species are endangered by environmental changes, such as increased UV levels*

pH

Organisms with exposed living tissues are particularly susceptible to enzyme damage caused by high or low pH.

Many plants can only survive if the soil is within a particular pH range. Outside this range their root cell enzymes may not function or they will be unable to absorb nutrients. For example, cowslips are often found on alkaline chalk soils and most lichens cannot survive acidic conditions.

Animals with **exoskeletons** cannot survive very acidic conditions as it dissolves the skeleton. For example, freshwater crustaceans such as crayfish cannot survive in acidic water.

Water

All organisms require water, but some species have particular requirements, often for breeding or because they do not have the ability to reduce water loss. For instance, frogs have thin skin that must remain moist. If it dries out then the living skin surface cells die and absorption of oxygen is reduced. Although they spend most of their lives out of water they live in moist areas and must return to water to lay their eggs. Toads have a thicker skin that reduces water loss so they can live in drier habitats.

Mineral nutrients

Animals get their mineral nutrients in their food or water, but most plants can only absorb them from their surroundings via their roots.

Case study

The sundew

Plants that live in anaerobic, waterlogged soils often have a shortage of nitrates. Some plants, such as the sundew, trap and dissolve insects to provide nitrogen compounds from their proteins. If the soil dries out it will become aerobic, nitrogen availability will increase and more vigorous plants such as grasses may out-compete the sundews.

Fig. 5.2 *The tough, flexible fronds of brown seaweeds protect them from storm damage*

Turbulence and physical damage

Some species are well adapted to surviving conditions of **turbulence** without being seriously damaged. For example, brown seaweeds on rocky shores are very flexible and are coated with mucus that reduces wear against rocks. Palm trees have long thin leaves with veins that run lengthways. If the leaves are split in storms the rest of the leaf can still survive.

Key terms

Turbulence: the roughness of the environment caused by chaotic air or water flow.

Species interdependence and control of abiotic factors

The abiotic factors that affect a species may be controlled or modified by other species living in the same habitat, so the survival of one organism may indirectly depend upon the presence of another species.

Altered abiotic factors in a woodland:

- light levels on the woodland floor may be lowered because of shading
- light absorption by chlorophyll in the canopy means woodland floor plants get little blue and red light, but more green light
- humidity will be higher because of transpiration by the canopy vegetation
- wind velocities will be reduced as trees act as windbreaks
- there will be an increase in nutrient availability because of the decomposition of dead leaves and branches.

Biotic factors

If a species is not adapted to the abiotic factors then it cannot survive. It must also be adapted to biotic factors, which involve other species, especially obtaining food and avoiding becoming food. Other biotic factors include disease, nutrient supply, and inter-relationships for breeding such as pollination and seed dispersal.

Feeding

All **heterotrophs** have to get their food energy from other living organisms. Depending on their particular food, they may need to have adaptations for catching, eating and digesting their food.

In harsh environments where food may be scarce or unreliable, it may be an advantage to eat a wide variety of different foods.

Tropical rainforests have abiotic conditions that are ideal for plant growth so productivity is high and food supplies are very abundant and reliable. This allows many animals to survive, with enough food for different species to exist with specialised feeding mechanisms to avoid competition. Birds' bills vary in shape to allow them to eat different types of food.

Table 5.1 *Feeding mechanisms requiring specific habitat protection*

Species	Specific habitat requirement
Parrots have strong, hooked bills to pick and open fruit and seeds; hummingbirds have long bills and tongues to feed on flower nectar	Habitat protection for these species must include enough suitable plants to provide fruit and nectar throughout the year
Most woodpeckers find insect larvae by pecking into rotten wood	Habitat management must include enough old rotting trees
The caterpillars of the large blue butterfly feed on wild thyme but spend the winter in the nests of a species of red ants where they eat ant larvae	Conservation of the butterfly must involve maintaining the grazed grassland so its food species thrive

Avoiding predators

The large number of herbivores and predators in rainforests make it important to have defence mechanisms such as a bad taste, toxins or thorns. Animals may also use camouflage to avoid being found.

Tree-living animals may be more prone to predation if they come down to the ground.

Fig. 5.3 *A rope bridge for howler monkeys in Belize, Central America*

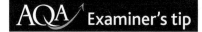 **Examiner's tip**

Concentrate on examples of species and adaptations where habitat management can help endangered species.

Key terms

Heterotroph: an organism that gains its organic compounds for energy and growth from other organisms.

Did you know?

Brown bears in Canada are omnivores and will eat roots, nuts, seeds, berries, honey, fruit, fish and carrion. This makes them easy to feed in captivity.

Did you know?

Nature reserves for howler monkeys in Central America and for dormice in the UK may use aerial links between trees so they can stay in the canopy.

Cellulase: an enzyme that digests cellulose that is produced by some bacteria, fungi and protozoans.

Cellulose: the carbohydrate made of linked glucose molecules, which is a major component of plant cell walls and wood.

Symbiosis: a relationship between organisms of different species that live together. One benefits while the other species may benefit, be unaffected or suffer.

Mycorrhizal fungi: symbiotic fungi associated with plant roots that gain carbohydrates from the plants and aid the uptake of nutrients such as phosphates from the soil by the plants.

Pollination: the transfer of the male plant gamets onto the female part of a flower, resulting in fertilisation and seed production.

Fig. 5.4 *Many flowers have evolved to attract particular pollinator species*

Fig. 5.5 *Many tree species rely on animals to spread their seeds*

Symbiotic nutrition

In some cases, different species have evolved to assist each other with nutrition. Some examples are as follows.

- The algae that live inside coral polyps on coral reefs are provided with a safe habitat and supplies of nutrients, while the polyps are provided with carbohydrates.

- No animals can produce the enzyme **cellulase** so they cannot digest **cellulose**, which is one of the most abundant food resources. All big herbivores, such as elephants, antelopes and cattle, have **symbiotic** cellulose-digesting bacteria in their intestines, as do many small herbivores, such as termites.

- Lady's slipper orchids have very small seeds, which spread easily but are not big enough to produce young plants that are big enough to be independent. The germinated seeds join with **mycorrhizal fungi** in the soil, receiving sugars, water and mineral nutrients. After several years the orchids produce leaves, start to photosynthesise and provide sugars to the fungi.

Pollination

Wind **pollination** is haphazard and is used mainly by plants that grow in dense groups such as grasses in a grassland or oak trees in an oak forest. Plants with a low population density where individuals are spaced well apart need a pollen delivery system, so they have evolved flowers that attract animals, such as bees, beetles, moths or bats that transfer pollen from flower to flower. Particular plants have evolved to attract specific animal groups using scent, colour, shape or time of opening.

Seed dispersal

Wind dispersal of seeds only works well for small seeds and plants in windy areas. Larger seeds cannot be carried by the wind and neither can those produced in sheltered areas such as beneath the forest canopy. Plants often produce seeds that are intended to attract animals. Some seeds may be carried away and eaten, but some may be dropped and can then grow into plants. For example, grey squirrels bury acorns as food stores for the winter. Many are never dug up again and will germinate to produce new oak trees. Some seeds are enclosed in edible fruit that attracts the animals. The seeds may be discarded or pass through the gut undigested. For example, tropical fruit bats eat fruit. The seeds are then dispersed in their faeces.

Disease

Diseases naturally control populations, especially when the population density is high. While individuals may become ill or die, the overall population may benefit by ensuring the weaker individuals do not survive, but if a population is already threatened then losses due to disease may cause it to decline.

Case study

Squirrel pox

Grey squirrels carry the disease squirrel pox. It does not kill them, but it does kill red squirrels. In areas where both squirrel species are found and the squirrel pox disease occurs, attempts are made to keep them apart. For example feeding stations are not used as they would increase the risk of both species meeting and the disease being passed from grey to red squirrels.

Nutrient supply

Nutrients from dead organic matter are released during the break-up by detritivores and subsequent breakdown by decomposers. Plants can then utilise the nutrients again, which ensures future food supplies for the detritivores and decomposers.

Indirect benefits

Species may be interdependent in many more subtle ways than the more obvious eat/be eaten link, pollination and seed dispersal. What a species does as part of its normal way of life may benefit other species. Some examples of indirect benefits are as follows.

- Many hole-nesting birds rely on the survival of woodpeckers to excavate the original holes.
- Forest elephants make large waterhole clearings called bais where they get water and mineral salts. They are also used by gorillas, deer, antelopes, many birds, insects, etc., but without the elephants the bais would grow over.
- Mangrove forests have roots that fix firmly into the sediments. The branches absorb the energy of storm waves and protect the other plants further inland. Conservation of the other species may rely on conservation of the mangroves.

To protect a particular species, it may be necessary to protect the complete **community** of species with which they interact.

Grouping organisms and environments

Organisms and their environments can be categorised in many ways that help us to understand their roles and interrelations.

Taxonomy

Taxonomy is the study of how organisms can be grouped according to how closely related they are, each group being called a **taxon**. Group types include kingdom, phylum, class, order, family, **genus** and species. These are largely subjective and depend upon opinions on the features that should be used for classification. The only group that is not subjective is species.

Species

A species is a group of organisms that resemble one another more closely than members of other groups and form a reproductively isolated group that naturally breed to produce fertile offspring.

Oak trees (*Quercus robur*) and house sparrows (*Passer domesticus*) are examples of species.

Most species could not produce fertile offspring with other species because they are genetically incompatible with different numbers of chromosomes or chromosomes that are different shapes and cannot line up with each other during fertilisation and cell division.

Very closely related species may be capable of producing fertile offspring where they are genetically very similar, but this does not happen in the wild, either because they live in different areas or because their behaviour prevents inter breeding, for instance because they have different courtship patterns. In captivity natural behavioural patterns can break down, which allows individuals of different species to mate.

AQA Examiner's tip

Many of these terms explained here are used in everyday language. Make sure you know their precise meanings so you can use them and interpret examination questions correctly.

Key terms

Community: a community of species refers to the populations of all the species living in a particular area.

Taxonomy: the study of organisms to assess how they may be grouped or classified.

Taxon: a group of organisms based on their biological similarities.

Genus: a group of closely related species.

Fig. 5.6 *This beach is suffering erosion as mangrove forests have been removed*

Hint

Make sure you can define the technical terms listed here and explain the difference between them. This will help you to describe other issues more clearly.

Did you know?

- Many duck species have evolved to become different species relatively recently so they are genetically similar. Natural courtship displays are very different, but in captivity they easily inter breed.
- One-humped Arabian camels and two-humped Bactrian camels are found in different parts of the world, but in captivity they can interbreed.

Species are given Latin names that are used worldwide to avoid confusion caused by the use of different languages or local names.

The first part relates to the genus, which is a group of very closely related species that have particular similarities that suggest they have diverged in evolution relatively recently.

The genus *Passer* includes 20 closely related sparrow species that are thought to be particularly closely related to each other.

The second part of the name is added to identify one species within the genus, thus *Passer domesticus* is the house sparrow of Europe, Asia and North Africa.

Population

Population refers to all the individuals of a species living in a particular area, for example all the oak trees in a deciduous woodland or all the carp in a lake.

Community

Community refers to the populations of all the species of plants and animals living in a particular area, for example all the oak trees, ash trees, holly bushes, hazel bushes, grasses, mosses, sparrow hawks, sparrows, blue tits, voles, earthworms, fungi, etc. in a woodland.

Grouping environments

Ecosystem

The **ecosystem** is the community of organisms living in an area, their inter-relationships and interactions with their abiotic environment; for example the community of species in a woodland, their inter-relationships and interactions with energy, water and nutrient cycles.

Physical processes such as energy and nutrient movements connect different ecosytems and some organisms will move between ecosystems, such as those organisms that migrate. However, ecosystems are usually considered as relatively self-contained functional units with almost all the species that interact with each other and their surroundings. Examples are coral reefs, rainforest, tundra, pond, estuary and savannah grassland.

Habitat

Habitat is the place where an organism, species or population lives. The habitat is usually a particular part of an ecosystem: a mole living in the soil within a grassland; blue tits living in the foliage of trees in a woodland; stag beetles living in the tree trunks of an oak woodland.

Niche

The niche of a species is the role it plays in its habitat. This includes the way it uses environmental resources, such as water, nutrients and food, and the way it changes its environment, such as reducing the availability of resources to other species or by providing food to its predators. For example, the tawny owl is a nocturnal carnivore, nesting in holes in hollow trees, feeding on small mammals and birds.

Two species cannot occupy exactly the same niche in the same place for any length of time. One species would be better adapted and would out-compete the other. The greater the degree of niche overlap, the more likely it is that one species will die out.

Key terms

Ecosystem: the community of organisms living in an area, their inter-relationships and interactions with their abiotic environment.

Activity

Create a table with columns to show details of four species' habitat, niche, biome and ecosystem.

Biome

A **biome** is a large geographical region with particular climatic features, in which a characteristic, unique community of species lives. A particular biome is all the areas of the world where a particular ecosystem type is found. They are usually named after the dominant organisms, usually the vegetation type. Examples include deciduous broadleaf woodland, coral reefs, temperate grasslands, mangroves and tundra.

Temperature, light levels, water availability and seasonal fluctuations in these are major climatic factors that control the distribution of biomes.

Biosphere

The biosphere is that part of the planet that is inhabited by living organisms: land surface, soil, water and atmosphere.

■ Changes in ecosystems

An understanding of natural temporal changes that occur in ecosystems helps in the understanding of the impact of human activities and therefore helps in developing effective conservation strategies.

Temporal changes: ecological succession

Primary succession

Primary succession involves the sequence of changes in community composition in an area, starting when it is colonised for the first time. The sequence of changing communities is called a sere and continues until the climax community develops. The climax community is usually controlled by the climate and is called the climatic climax.

The species that colonise the area change the abiotic factors and lead to more favourable conditions, which allow new species to colonise and out-compete the previous ones. As the abiotic conditions continue to become less extreme the diversity of species increases. Some of the newly arrived species can now survive because of the more moderate conditions, while others are more reliant on other species that are now present, for example for food.

Hydrosere

A hydrosere is a succession that begins in water. If an area of freshwater forms then it will soon be colonised by single celled algae from the soil. Herons, ducks and flying insects such as water beetles and pond skaters may also bring in spores and seeds of algae, diatoms and plants.

Rooted plants such as reeds, lilies and reedmace may colonise the lake edges but the open water may be too deep for rooted plants so only floating plants live there. As more plants colonise there is more food and shelter so more animals that arrive can survive.

As plants grow and die the lake gradually fills in. Soil and sediments may also be carried in from the surrounding area. The deep water becomes shallower until rooted plants can survive. Emergent plants that have their leaves above the water shade the submerged plants that gradually die out.

As the lake fills up with sediments open water disappears and aquatic species disappear. Trees such as willow and alder start to colonise as soil develops that can provide the support needed to hold up the weight of a tree.

The soil becomes drier as more sediments fill in and transpiration by trees removes water. Trees such as oak will eventually colonise. The denser

■ **Key terms**

Biome: a large geographical region with particular climatic features, in which a characteristic, unique community of species lives.

Ecological succession: the sequences of changes in community composition that changes as an area is colonised and develops until a climax community is eventually produced.

AQA Examiner's tip

Concentrate on understanding the processes involved in **ecological succession** and the changes in conditions rather than learning sequences of species.

Fig 5.7 *Semi-aquatic plants colonise the pond as it fills with sediments*

Fig. 5.8 *Wind dispersal of seeds is important when there are few animals to spread them*

canopy provides shade so smaller plants grow less well. Eventually, the community is dominated by the largest trees with other plants and animals living in, on or under them. Almost no species that were involved in the early stages are still present.

Animal species colonise as the conditions become suitable and die out when they become unsuitable. The speed with which they colonise depends upon the ease with which they can travel. Birds and flying insects arrive quickly, while animals like molluscs and worms come much later. If the area is isolated then some species may never arrive, even if the conditions are suitable.

Lithosere

A **lithosere** involves the development of the community on bare rock created by events such as a volcanic eruption, a cliff fall or the retreat of a glacier.

The early abiotic conditions are very harsh and unsuitable for most organisms. There is no soil; there are extremes of temperature and water availability. The first colonisers are simple **autotrophs**, such as algae and lichens. Weathered rock fragments and dead organic matter gradually build up. Mosses colonise and gradually a thin layer of soil builds up. Grasses and ferns colonise. As soil builds up and plants get larger the abiotic factors become less extreme. It never gets as hot or cold or wet or dry as when the rock was exposed. The soil makes plant nutrients available. Seedlings of less hardy plants can survive under the shade of the larger plants.

Flowering plants colonise and survive once the insects that pollinate them are established. When the soil is deep enough and the **edaphic** factors (soil factors) are suitable, trees can colonise. The first species usually have wind-blown seeds while later ones have seeds that are dispersed by animals.

In the early stages, a hydrosere and a lithosere are very different because of the different conditions. The final communities that develop are very similar and are mainly controlled by the climate of the region.

Areas that have water available all year usually become woodland. The temperature controls the type of woodland: tropical rainforest, temperate deciduous woodland or boreal conifer forest.

Areas with seasonal rain usually become grassland: tropical savannah or temperate grassland.

Secondary succession

Secondary succession refers to the changes that occur in an area that has already reached the climax state.

Natural events such as forest fires and hurricanes or human activities such as deforestation, grazing, mowing and burning interrupt the sequence of events in succession and can remove the climax community, possibly producing bare ground. Succession will start again but it will occur more rapidly than during primary succession because long periods of time are not needed for soil to develop as it already exists and many seeds are present in the soil. Plants and animals may also be able to colonise rapidly from surrounding undisturbed areas.

Plagioclimaxes

If the human activities that destroyed the climax community continue, then a new community of species will develop. This is called a

Fig. 5.9 *Wet meadows are maintained by grazing*

plagioclimax. If the human activity continues for long periods of time then the community may seem to be stable and it may not be obvious that it is only maintained because the natural process of secondary succession is being prevented from occurring.

If the human activity that produced a plagioclimax stops, then secondary succession will eventually re-establish the climax community. So, if the plagioclimax community includes species that are considered to be important then it is necessary to continue the activities that maintained the plagioclimax.

Conserving climax communities such as rainforest, coral reefs, mangroves and Antarctica usually involves the approach of minimal human impact, but in the UK conservation often involves management of plagioclimaxes and the continuance of previous activities.

Table 5.2 *Plagioclimaxes in the UK*

Habitat	Management practice
Lowland heathland	Grazing or burning
Hay meadow	Mowing
Wet meadow	Grazing
Upland moorland	Grazing or burning
Arable field	Ploughing
Garden lawn	Mowing
Coppiced woodland	Felling at intervals of 8 to 20 years
Reed beds	Mowing or cutting

Diversity and ecological stability

The ability to assess species diversity is important in monitoring environmental change, habitat damage and the success of conservation efforts.

Higher diversity in less abiotically extreme environments results in more stable ecosystems in which populations are dominated by biotic factors such as in tropical rainforests and coral reefs.

Diversity can be described quantitatively using a number of calculations. One of the most commonly used methods is Simpson's Diversity Index.

Estimates of the total number of species that exist

It is not easy to predict the number of things that have not yet been discovered, but predictions can be based upon the past rate of discovery. The gradual reduction in the ease with which new species can be found can be used to estimate the total number of species that exist.

Looking at areas that have not been thoroughly researched suggests that many species have yet to be discovered. Fewer than two million species have been named. Estimates of the total number of species that exist vary, but most suggest the total is between five and 100 million. Reasons why many species remain undiscovered include the following.

- Some ecosystems have not been thoroughly researched because they are too inaccessible, such as the deep sea floor and the canopy of tropical rainforests.

> **Link**
>
> For the Practical Skills needed in this unit, see Chapter 17.

> **Simpson's Diversity Index formula**
>
> $$D = \frac{N(N-1)}{\sum n(n-1)}$$
>
> where
>
> N = total number of organisms of all species and
>
> n = total number of organisms of a particular species
>
> $\sum$ = sum of
>
> The higher the calculated value, the higher the biodiversity.

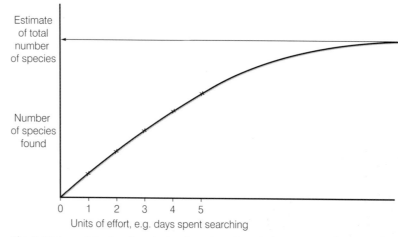

Fig. 5.10 *A graph to show how the total number of species in an area can be estimated*

Activity

Produce an information sheet of the potential value to humans of the species that have not been discovered yet.

Did you know?

Most birds start to breed in their first year, but albatrosses start to breed when they are quite old – up to the age of seven. They only lay one egg per year and may not breed every year.

■ Some species have such similar appearance, structures and behaviour that it is not possible to tell whether they are the same species or not.

■ In many species males and females look very different and live solitary lives, so it is difficult to tell they are the same species unless they mate.

■ Many plants can only be identified when they produce flowers.

New DNA and RNA identification methods are helping to identify species, to tell them apart and find how closely related they are.

Population dynamics

An understanding of the factors that can influence population change is important in monitoring species' survival, the success of conservation strategies and also in assessing the maximum sustainable yields of exploited species.

The actual size of a population is controlled by the numbers of births and deaths and movements in and out of the area:

$$\text{Population size} = \text{starting population} \begin{array}{l} + \text{births} + \text{immigrants} \\ - \text{deaths} - \text{emigrants} \end{array}$$

Factors affecting birth rates

The number of births is controlled by the natural reproductive potential of the species (biotic potential). This has evolved over long time periods to produce sufficient offspring to replace the individuals that die. There must be a surplus of young to ensure the survival of the population in bad years, without wasting time and resources that could threaten the survival of them all.

Species such as herring produce many young because the chances of dying are high. In contrast, species such as elephants produce few young because the chances of dying are low. These species can be particularly vulnerable to an increase in the death rates.

Factors affecting mortality rates

The number of deaths (mortality rate) is mainly controlled by environmental factors that prevent some of the individuals that are born from surviving.

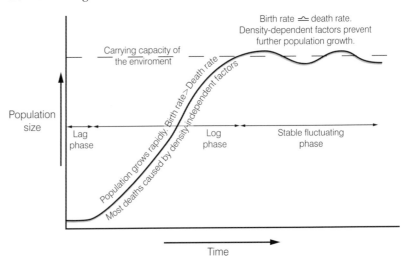

Fig. 5.11 *Sigmoidal population growth curve*

A range of environmental factors affects the likelihood of an organism dying: **density independent factors** include drought, flood and volcanic eruption; **density dependent factors** include food supply and disease.

Density dependent factors become more important as the population density increases, until the mortality rate forms a long-term balance with the birth rate.

Homeostatic regulation of population size

Carrying capacity

Carrying capacity is the maximum population size that can be supported indefinitely without damaging or over-exploiting the environment. If the population rises above the carrying capacity then density dependent factors become stronger, so mortality increases and the population decreases.

If the population drops below the carrying capacity then density dependent factors become weaker, so mortality decreases and the population increases. The mortality rate changes if the population is above or below the carrying capacity to return the population to the carrying capacity.

Predator: prey population relationships

This self-regulation of population can be shown in the predator:prey population relationships found in habitats where a prey species is the main food of a predator species. Where there are few food species available, a reduction in the availability of one food species will have a big impact on predator numbers.

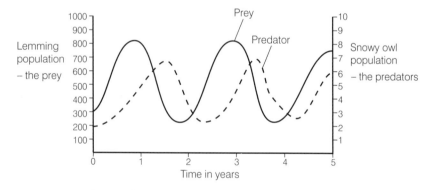

Fig. 5.12 *A predator:prey population relationship*

Artificial population control

Culling (selective killing) of a population may be necessary to conserve species or habitats where natural control mechanisms no longer exist, for example the culling of red deer in Scotland where wolves no longer exist.

■ **Key terms**

Density independent factors: a factor that is not influenced by the population density of the species that may be affected, e.g. drought and volcanic eruption.

Density dependent factors: a factor whose effect is affected by the density of the population, e.g. disease and food supply.

Carrying capacity: the greatest population that can be supported sustainably in an area.

Sigmoidal population growth: the growth pattern of a population where abundant resources allow rapid growth followed by population stabilisation as the carrying capacity is reached.

AQA Examiner's tip

Candidates should understand the factors controlling population growth in the lag, logarithmic and stable, fluctuating stages of the **sigmoidal population growth** curve.

Summary questions

1 Outline how an understanding of a named species' adaptations to abiotic factors helps with conservation management. *(4 marks)*

2 Outline how an understanding of a named species' adaptations to biotic factors helps with conservation management. *(4 marks)*

3 Explain how the successful conservation of species may depend on the conservation of other species. *(8 marks)*

4 Outline how some species change local abiotic factors, which influences the survival and distribution of other species. *(6 marks)*

5 Use named examples to define the following ecological terms: species, population, community, ecosystem, habitat, niche, biome, biosphere. *(3 marks each)*

6 Describe the changes in abiotic factors that occur during succession in a lithosere. *(6 marks)*

7 Explain how the changes in abiotic factors during a hydrosere control the changes in species that are present. *(6 marks)*

8 Explain how the species present cause the abiotic factors to change during a lithosere. *(6 marks)*

9 Explain how human activities maintain plagioclimaxes that provide important wildlife habitats. *(6 marks)*

10 Outline the factors that control the size of a population that can survive in an area. *(8 marks)*

6   Land resources

Learning objectives:

- How can important landscape be protected?

- How can decisions be made to resolve the conflicts between different possible uses?

Good management of the land surface of Earth is important. It is where we live, grow our food, spend most of our leisure time and collect most of our water. In some parts of the world the population density is so low that human impact is small and any impact is caused by larger problems such as global climate change. In countries such as Britain, the population density is high and human impacts are great. Many areas have multiple uses with several activities such as agriculture, wildlife conservation and public recreation all taking place together. In this situation a range of strategies to restrict, encourage and manage human activities must be used.

Landscape conservation for informal public recreation

Increased affluence, leisure time and car ownership have all made it easier for the public to travel from the urban areas where most people live into the countryside. As the public spend more time in the countryside, it is increasingly important to conserve the landscape so it can be enjoyed and used for informal public recreation. Since communities of plants and associated animals are an important part of the landscape, landscape conservation often results in wildlife conservation.

Landscape protection

Conserving the aesthetic appeal of the environment involves the maintenance of features that are natural or have been produced by human activities and give the countryside its character, for example woodlands, hedgerows, stone walls, in-field trees, ditches, banks, ponds and river features. Some of these features can be protected simply by preventing any damaging activities. Others need active management to counteract natural processes such as ecological succession.

Landscape enhancement

Restoration and development of countryside features can enhance its character such as planting small woodland areas, replacing conifers with mixed indigenous species, 'soft' riverbank management and the restoration of river meanders.

Visitor management

This should involve the careful provision of facilities that do not damage the character of the countryside, for example paths, waymarks, car parks, information and recreational facilities. They can be designed to fit in to their surroundings by the use of traditional designs and local materials.

Paths may be made of sand, gravel or stone. Buildings and signs may be made of rough wood. Car parks can be divided up with banks and hedges to disguise their size and the use of grass with protective matting produces a more natural surface than tarmac.

Fig. 6.1 *This footpath allows public access but also keeps people away from sensitive wildlife areas. It is also wheelchair accessible*

■ **Key terms**

Honeypot site: an area that is particularly attractive to visitors.

Countryside Council for Wales: the equivalent organisation in Wales to Natural England.

National Parks: designated areas for informal public recreation, wildlife conservation and maintenance of the rural economy.

National Park Authority (NPA): the organisation that runs a National Park.

■ **Link**

For more on the work of government organisations see Chapter 3.

Providing good facilities and publicising them will attract the public and help keep them away from sites where their presence would cause problems. These are often called **Honeypot sites**. They help protect sensitive wildlife sites and areas where visitors would cause congestion and disturbance to local residents. The extra facilities provided could include a visitors' centre, toilets, café, wardens, guided walks and other events.

Governmental organisations

Natural England/ Countryside Council for Wales

The organisations Natural England and **Countryside Council for Wales** are responsible for designating areas for landscape conservation. They set up controls and regulations which must be followed by landowners and other users to protect them from undesirable change. Damaging changes such as the clearance of natural woodland or urban expansion are prevented. Beneficial management activities are carried out such as the grazing of meadows and moorland to stop ecological succession and maintain the plagioclimax.

National Parks

The original 10 **National Parks** in the UK were established in the 1950s following the National Parks and Access to the Countryside Act (1949), which was a response to the growing demand for public access to scenically beautiful open areas of countryside. Other National Parks have been established more recently.

National Parks in the UK are very large areas, some covering more than 2,000 km². They are designated by Natural England, but each National Park is managed by its own **National Park Authority (NPA)**.

The aims of National Parks are:

■ to conserve and enhance their natural beauty, wildlife and cultural heritage

■ to promote opportunities for the understanding and enjoyment of their special qualities

■ to maintain the rural economy.

Fig. 6.2 *The wide open spaces of Dartmoor attract many visitors*

Table 6.1 *Land ownership in selected National Parks*

National Park	Land ownership /%						
	National Park Authority	Forestry Commission	National Trust	Natural England	Ministry of Defence	Water company	Others (mainly private owners)
Dartmoor	1.4	1.8	3.7	–	14.0	3.8	75.3
Exmoor	4.3	2.0	10.0	–	–	0.6	83.1
Snowdonia	0.5	15.6	8.7	–	–	6.0	69.2
Peak District	4.2	0.5	9.7	–	–	13.0	72.6
Pembrokeshire Coast	0.5	1.2	4.7	0.5	4.5	–	88.6

Because they do not own the National Parks, the National Park Authorities have to manage them by controlling the activities of the landowners through planning and development restrictions and management agreements. Consultation and discussion between all interested parties is very important.

Despite the aims of the National Parks and their protected status, some controversial developments have taken place that seem to contradict the aims of the National Parks. These are usually justified as being for the 'greater national good', which must override local needs.

There have been some controversial developments in and proposals for National Parks:

- military training, for example exercises on Dartmoor or pilot training over the Lake District
- reservoirs, for instance on Dartmoor
- quarrying for china clay in Dartmoor and limestone in the Peak District
- conifer plantations
- tourism developments
- energy, for example nuclear power stations, hydro electric power (HEP) stations and wind farms.

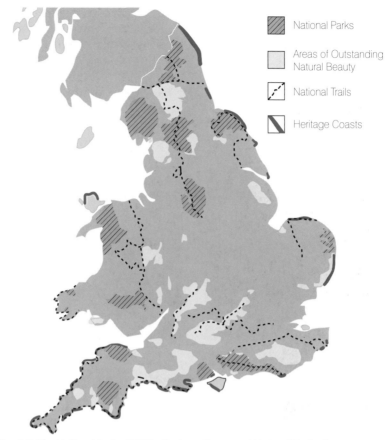

National Parks

Areas of Outstanding Natural Beauty

National Trails

Heritage Coasts

Fig. 6.3 *The National Parks, AONBs, Heritage Coasts and National Trails of England and Wales*

There are also several conflicts of interest and competing of interests within the National Parks as follows:

- Erosion – large numbers of people walking across the countryside can cause serious soil erosion. Cycling and horse riding can add to the problem. It may be necessary to restrict erosion damage by directing visitors away from vulnerable areas. Resistant surfaces such as paths made of gravel, wood or rock may reduce erosion problems.
- Congestion – large numbers of visitors in cars cause traffic congestion and increase pollution. Some National Parks encourage the use of buses, trains and bicycles instead of cars.
- Disturbance of wildlife – visitors can easily trample vegetation and disturb ground-nesting birds. Repeated disturbance can cause breeding failure. Access to such areas may be restricted or visitors may be encouraged to go to other areas.
- Litter – can cause livestock and wildlife to choke or become entangled and injured. Broken glass can act as a magnifying glass on sunny days and start fires.
- Displacement of the local community – the attractiveness of living in a scenically beautiful area has attracted people from outside the National Parks to buy houses for retirement or as holiday homes.

Fig. 6.4 *Footpaths make people less likely to walk anywhere at will across the countryside but by funnelling walkers the existence of footpaths can cause erosion, especially if not maintained*

This reduces the number of houses available for local residents and forces prices up. Restrictions on developments prevent lots of new houses being built to meet the demand. In Exmoor National Park some new homes are built specifically for locals.

- Conflicts between recreational users – quiet recreational activities such as walking and canoeing can be spoilt by more obtrusive activities such as water skiing or helicopter flights. In the Lake District National Park there is a 10 mph speed limit for boats, which prevents speedboats from causing annoyance or disturbance to other visitors.

Fig. 6.5 *Cycling and walking on the same path can cause problems*

Fig. 6.6 *Cycling is banned on many footpaths*

Fig. 6.7 *Public transport, cycling and walking are encouraged to reduce traffic congestion*

Activity

Produce a table to show all the different types of area designated in the UK for landscape conservation.

Include the aims, organisations involved, important legislation or agreements and examples.

Your aim is to help you to distinguish between them by identifying their main differences.

Areas of Outstanding Natural Beauty

AONBs are designated for their landscape qualities for the purpose of conserving and enhancing their natural beauty (which includes landform and geology, plants and animals, landscape features and the rich history of human settlement over the centuries). They are designated by Natural England but administered by the relevant County Council.

Like the National Parks, these are areas of scenic and recreational amenity value. They are generally less wild and remote and are usually hilly or lowland areas. They are often more intensively used than National Parks so economic activities such as agriculture, rural industry and residential areas are more important. It would be inappropriate to place strict restrictions on such activities.

There are generally fewer opportunities for open air recreation in AONBs than in National Parks and rights of public access are not an aim, unlike in National Parks.

Proposals for new developments are controlled by planning regulations, which are more restrictive than in most parts of the country, but less strict than in National Parks. Housing developments may be permitted, but often only to satisfy local housing needs. Expansion of an existing local industry may be permitted but the development of a new industry may not. Examples of AONBs are:

- Forest of Bowland
- North Pennines
- Cannock Chase
- Cotswolds
- Scilly Isles.

Heritage Coasts

Heritage Coasts are designated by Natural England as the finest stretches of undeveloped coastline that need to be protected from development for their scenic amenity value and, where appropriate, public access may be increased.

Some Heritage Coasts in England are:

- North Northumberland
- Flamborough Headland
- South Devon
- St Bees Head.

National Trails or Long-Distance Footpaths

National Trails or **Long-Distance Footpaths** are designated by Natural England to establish public rights of way through landscape of great scenic value. They have been created by linking footpaths, bridleways and minor roads to produce routes for walkers, cyclists and horse riders. National Trails in England include:

- Cleveland Way
- Pennine Way
- South Downs Way
- South West Coast Path.

DEFRA (Department for Environment, Food and Rural Affairs)

Agri-environmental schemes set up under DEFRA, such as the Environmental Stewardship Scheme (ESS), could potentially influence all farmland and therefore have a big impact on the aesthetic appeal of the countryside. Some aspects are intended to maintain historically important features of the landscape such as dry stone walls and archaeological features. Other features are mainly intended to protect wildlife but also affect the appearance of the landscape, such as hedgerows and in-field trees. The higher level scheme also includes footpaths and educational access. (For more on the ESS, see page 256.)

Local authorities (and others)

Country parks

Country parks are areas of land, usually in or near urban centres, intended to provide informal recreational opportunities for the public. Many are abandoned or redeveloped derelict industrial sites or have other uses such as plantations or reservoirs.

Urban green space

Green spaces in urban areas or **urban parks** can provide a valuable facility for people that may rarely see the countryside. The issues of visitor and habitat management apply here as much as in rural habitats.

Non-governmental organisations

The National Trust

The National Trust was set up over 100 years ago to protect threatened coastline, countryside and buildings from uncontrolled development for public enjoyment.

Key terms

Heritage Coast: a coastline protected from development for its scenic or environmental value.

National Trails: long-distance routes for walking, cycling and horse-riding.

Long-Distance Footpaths: long-distance routes for walkers, most of which are also National Trails.

Country park: an area of countryside managed for public enjoyment. Most are near urban areas and are run by local authorities.

Urban park: an area of semi-natural land in an urban area used for public recreation and relaxation.

Summary questions

1. Outline the roles of the following governmental organisations involved in landscape conservation in the UK:
   - Natural England/Countryside Council for Wales *(6 marks)*
   - Local authorities *(4 marks)*
   - DEFRA. *(4 marks)*

2. Outline the functions of the different designations that protect landscapes in the UK. *(6 marks)*

The **Neptune Coastline Campaign** involves the purchase and protection of important coastlines.

In addition to protected historically important buildings, the National Trust protects over 1,130 km (700 miles) of coastline and about 250,000 hectares of countryside, including forests, farmland and moorland. Among the areas protected by the National Trust are:

- Farne Islands, Northumberland
- Malham Tarn, North Yorkshire
- Wicken Fen National Nature Reserve, Cambridgeshire
- White Cliffs of Dover, Kent
- Lundy Island, Devon.

Important habitats are managed or created for wildlife, often using traditional techniques, including organic farming. They also encourage the use of public transport and sell local produce in their cafes.

Other NGOs involved in landscape conservation include the Campaign to Protect Rural England (CPRE) and organisations such as the RSPB and the Wildfowl and Wetlands Trust whose primary role is wildlife conservation.

■ Land-use conflicts

The competition for land resources in the UK leads to conflicts involving proposed, adjacent or multiple land uses and their possible or actual impacts upon the environment.

Major causes of land-use conflicts

Urban expansion

Changes in population size, social structure and the attraction of areas of prosperity have caused an increase in demand for housing land, especially in the South East of England. Where possible, brownfield sites are used such as those where old industries have declined. Many housing developments are on urban green spaces or where single buildings are replaced with multiple dwellings. This can increase overcrowding and congestion and reduce the quality of life within the urban area. The alternative may be to build on greenfield sites such as farmland around the urban area or in new or expanded towns.

Transport developments

Road schemes

Road usage in the UK has increased greatly in the last 50 years. People travel more, especially by car. As society has become more affluent more goods are bought, which has increased the transport of raw materials and manufactured goods. Much of this extra transport involves roads.

The congestion caused by the extra cars and lorries has led to many schemes to expand existing roads or build new ones. Some controversial road schemes include:

- M3 extension at Twyford Down
- A34 Newbury bypass
- A66 Temple Somerby bypass
- M6 toll road.

Fig. 6.8 *The M3 at Twyford Down destroyed important wildlife habitats*

Proposed or enlarged airports

The growth in air travel has required the expansion of airports which are usually near urban centres. The airports result in the loss of farmland, natural habitats and rural communities. Noise and congestion problems cover a much larger area. Some controversial airport developments are:

■ Heathrow Terminal 5, proposed third runway and Terminal 6

■ Manchester airport second runway

■ expansion of Stansted airport.

Port developments

Much of the increased trade in and out of the UK is carried by sea. Ports have to be in sheltered locations such as estuaries, which involve the destruction of sensitive habitats such as mudflats and saltmarsh. There have been several controversial port developments including:

■ container port at Dibden Bay in Southampton Water

■ container port at Harwich in Essex

■ oil terminals at Milford Haven in Wales.

Mining and quarrying

Minerals can only be extracted where they are found in sufficiently large deposits. Alluvial deposits such as sand, gravel and clay are important for building construction. They are often extracted in lowland areas where farmland is often more fertile and the population density is high.

Some minerals are found in upland areas, such as china clay and granite in Dartmoor and limestone in the Peak District.

Case study

The Glensanda super-quarry

The Glensanda super-quarry on the west coast of Scotland has produced huge amounts of granite for roads, airport runways and to line the Channel Tunnel. The coastal location means the rock can be exported by ship not by road. Having one huge quarry focuses the environmental impact on one area rather than spreading it over a range of sites.

Fig. 6.9 *Many china clay quarries are in upland areas of scenic value*

Harnessing energy

Wind farms

The windiest places are often the most scenically attractive, such as uplands and coastal areas. Many people think wind farms spoil the view. The 'swoosh' of the blades, hum from the generator and flicker of sunlight reflected from blades can be a problem for people living close to the wind farm.

Fig. 6.10 *Wind farms in upland areas harness stronger winds but may cause more aesthetic damage*

Habitat loss is caused by the tower foundations and access paths. The rotating blades may kill birds, especially for species that fly a lot for hunting or defending territories, such as eagles or lapwings. Deaths are also more likely if the wind farms are located on important migration routes, such as those for geese or swans.

Sensitive wind farm location avoids areas of great scenic beauty, where noise or radio interference may cause problems and where bird and bat strikes are likely.

Hydro electric power (HEP) stations

HEP stations are usually located in upland areas of scenic beauty.

Case study

Dinorwig HEP station

Dinorwig HEP station was built within Snowdonia National Park in the 1980s. The environmental impacts were reduced in several ways:

- the turbine and generator rooms were built underground in excavated chambers
- the pipes between the reservoirs were tunnelled underground
- the development was landscaped using waste from derelict slate quarries
- the power lines linking with the national grid are underground, so no pylons are needed.

Tidal barrages

The flow of water in and out of estuaries with the tidal cycle gives the opportunity to generate a lot of electricity, but the impact on sensitive estuary habitats would be great due to changes in water level, currents and turbidity. Shipping would also be affected and the materials needed to construct the barrage would need to be extracted by quarrying.

All these schemes have damaging impacts, but if they are not used then the energy will have to be produced another way, which will have different impacts, possibly worse ones.

Recreation and tourism

Increasing numbers of people visiting the countryside to enjoy it threaten to damage it by increased congestion, trampling damage and the damage caused by providing the facilities they require such as car parks. The activities of different groups of visitors may conflict with each other and need to be regulated.

Waste disposal

The wastes produced by society have to go somewhere, but all of the options have problems. Landfill sites take up land and release methane, incinerators are unpopular with locals and the resource value of the waste is lost. Recycling would be much better but it requires public cooperation.

Activity

Find a major development that is proposed or is under development. Produce an information sheet that shows details of the likely environmental impacts and ways in which they could be reduced.

Summary questions

3 Describe how the following human activities cause land-use conflicts in the UK:

- Urban expansion
- Transport developments
- Mining and quarrying
- Harnessing energy
- Recreation and tourism
- Waste disposal.
  (3 marks each)

Human impacts on the atmosphere

The composition of the atmosphere is being changed by human activities.

Many of the processes that affect the atmosphere are interconnected. So, if one process is changed, it may alter some of the others. It is important to understand this, because human actions may change one process that then causes others to alter as a consequence of the first change. We do not understand everything about the atmospheric processes so we cannot predict accurately how a human activity may affect the atmosphere.

Many of these human impacts are covered in Chapter 11.

Global climate change

The greenhouse effect

The **greenhouse effect** is the natural phenomenon that controls the energy and heat balance of the Earth's surface and the atmosphere. Visible light from the sun passes through the atmosphere relatively easily. The light that is absorbed by the Earth is converted to heat and then emitted as infrared energy or radiation. The infrared radiation that is emitted by the Earth's surface is partly absorbed by gases in the atmosphere, which causes the atmosphere to become warmer until the surplus heat energy is eventually radiated into space as infrared radiation.

The enhanced greenhouse effect and global climate change

Human activities are increasing the concentration of gases that absorb infrared radiation, which is causing the atmosphere to warm up. This is the **enhanced greenhouse effect**. Some of these are gases that are naturally found in the atmosphere while others are only released by human activities.

Table 7.2 *The major anthropogenic sources of greenhouse gases*

Greenhouse gas	Human activities	Approximate relative effect per molecule
Carbon dioxide	Combustion of fossil fuels, deforestation	1
Methane	Anaerobic bacteria in paddy fields, landfill sites and the intestines of livestock. Methane is produced during the formation of fossil fuels and released by the ventilation of coal mines, leaks from natural gas fields and pipelines	25
Oxides of nitrogen	Oxygen and nitrogen from the air react at high temperatures in vehicle engines and power stations. They are then released into the atmosphere in the exhaust gases	2,000
Chlorofluorocarbons (CFCs)	CFCs were used as aerosol propellants, fire extinguishers, refrigerants, solvents, and in expanded foam plastics	5,600 (CFC-11)
Tropospheric ozone	Produced by the photochemical breakdown of NO_2 and subsequent reaction with oxygen	25,000

The likely consequences of global climate change

The changes in the atmosphere cause it to retain more heat energy and warm up, but the actual temperature changes are quite small and are less than the temperature fluctuations during a typical day. The mean global temperature has risen about 0.75°C in the last 100 years. Accurate predictions are difficult, but a rise of one to 4°C in the next 100 years seems likely. Even this temperature rise would have relatively small impacts compared with the indirect effects of changes in other natural processes.

Sea level rise

A temperature rise will cause sea level to rise in two ways: thermal expansion and melting land ice.

Thermal expansion

The warmer atmosphere causes the sea to warm up and therefore expand, causing sea level to rise. This will be a very slow process because there is an enormous amount of water in the oceans. Water has a high specific heat capacity so it will take a long time for the temperature of the sea to 'catch up' with the warmer atmosphere and for all the expansion to take place. Only the water at the surface is warmed by direct contact with the atmosphere, so deep water will only warm up when the slow-moving ocean currents bring it to the surface.

Melting land ice

As the Earth warms up ice will melt. Ice that is floating on the sea does not cause sea level to rise when it melts as it contracts during melting and occupies the same volume as the volume of water it displaced when it was ice.

Ice that is on land will cause sea level to rise as the water flowing into the sea increases the volume of water in the sea. Glaciers and Antarctic ice shelves form on land and cause sea level to rise as they flow off the land and displace seawater.

The mean rate of sea level rise over the last 100 years has been 1.7 mm per year, mainly due to thermal expansion and the melting of some glaciers.

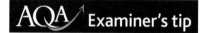

Fig. 7.5 *Many low-lying islands and coastal areas would be flooded by a rise in sea level*

■ Different changes may occur in different locations or at different times. If one area gets less rain then another area may get more. If Britain warms up then there may be less rainfall in summer as the water vapour does not condense, but in winter it is cold enough for condensation, and higher evaporation rates over the ocean means that more rain falls. So, simple questions such as 'Will it rain more?' have no simple answer.

■ We do not fully understand how all the Earth's climate systems work so we cannot predict the consequences of a particular human action.

Feedback mechanisms

A change in one environmental factor may cause other features to change. These may have an effect on the original change, possibly increasing it or maybe reducing it.

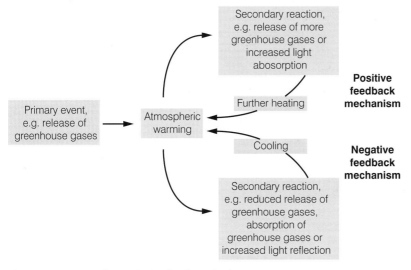

Fig. 7.14 *Positive and negative feedback mechanisms*

Fig. 7.15 *Forest fires are a natural phenomenon but they may become more common*

Positive feedback mechanisms

A **positive feedback mechanism** occurs where an environmental change causes other changes that increase the rate or amount of the original change and therefore increase its effect. These may increase temperatures directly or increase the concentrations of the gases that will cause further temperature rise.

Raised temperatures may cause the following.

■ Warming may increase the rate of decomposition, causing more carbon dioxide to be released and therefore more warming.

■ Warming reduces the area of ice and snow on the ground or floating on the sea. This reduces the albedo so less sunlight is reflected, solar heating is increased and temperatures rise further.

■ Warming of the sea may eventually cause methane to be released from methane hydrate in marine sediments. This would cause further warming.

■ Melting permafrost in polar regions releases methane gas bubbles that were trapped in the ice. This causes further heating.

■ Warming may cause forests and areas with peat soils to become drier. Fires may become more frequent and last longer, adding extra carbon dioxide to the atmosphere.

■ **Key terms**

Positive feedback mechanism: a situation where an initial change causes a reaction that increases the original change.

Negative feedback mechanisms

A **negative feedback mechanism** occurs where an environmental change causes other changes that decrease the rate or amount of the initial change and therefore reduce its effect and help to re-establish the original equilibrium.

■ Higher temperatures may cause increased rates of photosynthesis, which would store more carbon in biomass. Levels of carbon dioxide in the atmosphere would be lowered, which would cause cooling.

■ Warming would increase evaporation and cause the formation of more low-level cloud. This would increase the albedo, reflect away sunlight and reduce solar heating.

Fig. 7.16 *Low-level cloud reflects sunlight and reduces the amount of energy absorbed and converted to heat*

Strategies to control global climate change

Table 7.3 *Methods of reducing greenhouse gas levels*

Greenhouse gas	Method
Carbon dioxide	Energy conservation to reduce use of fossil fuels Use of alternative energy resources Carbon sequestration by planting more trees or storage of carbon dioxide from power stations underground **Kyoto Protocol** emission reductions
Methane	Reduced dumping of waste in landfill sites Reduced livestock production Better collection of gas from coal mines and gas and oil facilities
Oxides of nitrogen	Reduced use of internal combustion, such as more use of public transport Catalytic converters in vehicle exhausts Addition of urea to power station effluents or diesel engine exhausts
Chlorofluorocarbons	Use of alternative materials: – butane or propane in aerosol cans – hydrofluorocarbons (**HFCs**) and hydrochlorofluorocarbons (**HCFCs**) in refrigerators – alcohols as solvents for cleaning electronic equipment Use of alternative processes: – trigger and pump action spray cleaners instead of aerosol cans – stick deodorants instead of aerosol cans
Tropospheric ozone	Methods that reduce NO_x emissions reduce the formation of tropospheric ozone

Case study

The Kyoto Protocol (1997)

MEDCs that signed up to the treaty were legally bound to reduce their emissions of six greenhouse gases (collectively) by an average of 5.2 per cent below their 1990 levels by the period 2008–12. Failure to achieve this leads to their permitted emissions for the next period being cut by a further 30 per cent.

Each country has developed its own method to meet its target. The EU has identified 12,000 factories and power stations that have been given a carbon dioxide quota. If they exceed this they can purchase extra allowances or pay a financial penalty. If they fall below the amount they can sell the extra quota.

LEDCs do not have legally binding emission limits as this would unfairly hinder their development.

Agreements such as the Kyoto Protocol do not reduce emissions themselves, but they encourage the use of methods that will.

Fig. 7.17 *Temporary flood barriers to prevent river flooding*

The concentrations of most **anthropogenic** greenhouse gases continue to rise as the activities that release them have not been successfully controlled. The exception is the **CFCs** that are being phased out to protect the stratospheric ozone layer.

Strategies to cope with climate change

Climate change may require changes in lifestyle, infrastructure services and land use.

Key terms

MEDCs: More Economically Developed Countries.

Anthropogenic: something made or caused by human activities.

CFCs: chlorofluorocarbons are pollutants that cause ozone depletion and contribute to global climate change.

Infiltration: the process by which surface water enters the ground between the particles of soil or rock.

Table 7.4 *Strategies to cope with climate change*

Activity or problem	Strategies
Agriculture	Cultivate warmer climate crops Cultivate drought-resistant crops Abandon areas where irrigation is unsustainable Increase soil organic matter to increase water retention Water storage in times of water surplus for later irrigation use
Building design	Better ventilation and cooling systems to reduce the use of air conditioning Paler materials to reduce heat absorption
Flooding	Riverbank defences River barrages to protect against high tides Less building on flood plains Reduce runoff rates, for example, by reducing paved areas to increase **infiltration** River regulation dams
Coastal erosion	Improved coastal defences Managed retreat – abandon lower value areas that are difficult to defend
Storm damage	Design of stronger buildings

Ozone depletion

Since the 1970s a reduction in atmospheric ozone has been observed in two different ways.

■ A general global reduction of about 4 per cent per decade over the whole planet.

■ So-called 'ozone holes' in polar regions where there is an annual drop in stratospheric ozone concentrations. Concentrations over Antarctica have dropped by up to two-thirds while over the Arctic the drop has been up to one-third.

Why is there concern about ozone depletion?

If UVB is not absorbed in the atmosphere then it will reach the Earth's surface and may be absorbed by living cells. The energy of the UVB is absorbed and some is converted to chemical energy as it breaks up biological molecules causing damage to the DNA in exposed cells, skin cancer, cataracts and damage to plant tissue and marine plankton.

Ultraviolet (UV) light absorption

Ultraviolet (UV) light is categorised by its wavelength, as shown in the table.

Table 7.5 Categories of UV light

Type of UV light	Wavelength	Characteristics
UV A	320–400 nm	Not absorbed by ozone
UV B	280–320 nm	Almost fully absorbed by ozone
UV C	<280 nm	Completely absorbed by ozone and normal oxygen

The UV light considered in studying ozone depletion is UVB.

The effects of UVB light on the gases in the atmosphere

UVB is involved in a number of chemical reactions. It is absorbed in providing the energy for the reactions so very little reaches the Earth's surface.

UVB is absorbed by both diatomic and triatomic oxygen in photolytic reactions where the molecules are split. The products of these reactions are involved in further reactions.

Summary of reactions

$$O_3 \underset{UV}{\overset{UV}{\rightleftarrows}} O_2 + O$$

Chlorine and ozone depletion

Halogens, especially chlorine in the stratosphere, cause ozone depletion in a range of reactions.

A single chlorine atom is a chlorine radical because it has an unpaired electron. This makes it very reactive.

A chlorine radical can react with an ozone molecule, breaking it down to leave an oxygen molecule and chlorine monoxide.

$$Cl + O_3 \rightarrow ClO + O_2$$

The chlorine monoxide can then react with a monatomic oxygen atom to make chlorine dioxide.

$$ClO + O \rightarrow ClO_2$$

The ClO_2 can then break down to release the chlorine again which may repeat these reactions thousands of times.

$$ClO_2 \rightarrow Cl + O_2$$

So, the reactions start and end with a free chlorine radical.

The important overall summary reaction is:

$$Cl + O_3 + O \rightarrow 2O_2 + Cl$$

This therefore removes important starting materials for both parts of the reactions in the dynamic equilibrium that maintains the ozone layer.

How does chlorine get to the stratosphere?

Chlorine is very reactive so if it is released in the troposphere it will probably react with something else and never reach the stratosphere.

Pollution by chlorofluorocarbons

Chlorofluorocarbons (CFCs) contain only chlorine, fluorine and carbon. They were developed as the fluids for refrigerators in the 1920s and were later used as aerosol propellants, solvents and as the gases for expanded plastics such as expanded polystyrene.

CFCs have properties that made them suitable for many uses. They are chemically very stable, can be liquefied with pressure, are good solvents of grease and oils, are not flammable and most are not toxic.

Their chemical stability and low solubility in water mean that they can remain in the atmosphere for years if they are released so they may eventually get carried up into the stratosphere. CFCs do not reduce ozone depletion but they are broken down by UV light in the stratosphere to release chlorine, which then causes the damage.

There are many CFCs; this example reaction uses CFC11.

$$CFCl_3 + UV\ light \rightarrow CFCl_2 + Cl$$

CFC emission reduction

The **Montreal Protocol** (1987) is an international agreement, which phased out the manufacture and use of CFCs and other ozone-depleting substances such as bromine-containing fire extinguisher propellants and 1,1,1 trichloroethane, which was a widely used solvent.

A range of alternative materials has been used to phase out CFCs, for example alcohols and HCFCs. Alternative processes include trigger pack sprays. (See Table 7.3 for more on this.)

CFCs have been replaced in refrigerators with HCFCs. These also contain chlorine but they are less chemically stable so they break down before they reach the stratosphere and thus cannot damage the ozone layer.

Hint

Understanding the reactions involved in ozone depletion is important. But it is more important to understand the consequences: these are that if they occur then less UV will be absorbed and less will reach the Earth's surface.

Fig. 7.18 *Before ozone depletion was detected, there seemed to be no disadvantages to using CFCs. This can held CFCs that were used to clean computer electronics in a hospital. On the back it states 'Environmentally safe propellant' – referring to their non-toxicity and worker safety*

Key terms

Montreal Protocol: international agreement that has controlled the release of ozone-depleting substances.

Why is ozone depletion greatest in Polar Regions?

The chemical reactions involving chlorine and ozone require UV B but occur most easily at very low temperatures when ice crystals form that provide catalytic surfaces. During the Polar winter the temperatures drop and allow the ice crystals to form, but there is no sunlight. When spring arrives, the crystals still exist and the sunlight causes the ozone-depleting reactions to take place.

Do other gases cause ozone depletion?

Oxides of nitrogen have been identified as possible causes of ozone depletion but releases in the troposphere do not reach the stratosphere as they dissolve in rain or react with other substances. High flying aircraft do release NO_x into the stratosphere, but there are relatively very few, so no obvious damage is caused.

■ Hint

- Make sure you are clear about the differences between global climate change and ozone depletion.
- Chlorofluorocarbons cause both problems, but in very different ways.
- Ozone in the troposphere is produced by human activities and is a cause of global climate change. Its presence is the problem.
- Ozone in the stratosphere is naturally present and is damaged by human activities. Its absence is the problem.

■ Summary questions

1. Outline the importance of the gases in the atmosphere for life on Earth. *(8 marks)*

2. Explain how life on Earth has changed the composition of the atmosphere. *(4 marks)*

3. Outline the role of the atmosphere in protecting the Earth's surface from UV light. *(3 marks)*

4. Outline the processes involved in the natural Greenhouse Effect. *(5 marks)*

5. Explain how human activities have altered the Greenhouse Effect. *(8 marks)*

6. Describe the likely effects of global climate change on:
   - sea level *(4 marks)*
   - ocean currents *(4 marks)*
   - The survival and distribution of wildlife. *(10 marks)*

7. Explain why it is difficult to accurately predict the effects of global climate change. *(8 marks)*

8. Describe how positive feedback mechanisms may increase the effects of global climate change. *(6 marks)*

9. Describe the methods that can be used to reduce global climate change. *(10 marks)*

10. Explain how human activities have caused ozone depletion. *(4 marks)*

11. Outline the methods that have been used to reduce ozone depletion. *(6 marks)*

8   The hydrosphere

Learning objectives:

- What are the unusual properties of water and how do these properties control the behaviour of water on Earth?

- What are the uses of water, and how can it be exploited?

- What problems may this exploitation cause and what solutions may be available?

Key terms

Evaporation: the change of water from liquid to gas as hydrogen bonds are broken.

Hydrogen bond: the weak electrostatic bond formed between water molecules, which gives water a high boiling point.

Anomalous expansion: unusual expansion, usually applied to water, which expands as it is cooled below 4 °C.

The hydrosphere contains water in all its forms (solid, liquid and vapour) that are found on, in and around the Earth.

The properties of water

Changes of state

The shape of a water molecule and the way the electrons are arranged allows the negative parts of one molecule to form weak bonds with the positive parts of others.

These bonds produce groups of up to four molecules that behave like a larger molecule with a higher boiling point. So, water is a liquid at temperatures when its low molecular mass would normally make it a gas. The bonds are continually forming and being broken. Individual water molecules at the water surface may escape as a gas: **evaporation**. At 100 °C and one atmosphere pressure all **hydrogen bonds** are broken and the water turns to gas as it boils.

The narrow temperature range within which water changes state between solid, liquid and gas allows the hydrological cycle to occur.

Anomalous expansion

The water molecules in ice are less densely packed than when they are in liquid form so solid ice floats on liquid water.

In an area that has cold winters, the water in a lake will be cooled by the cold air above. As the water is cooled below 4 °C, the water molecules start to take up the arrangement they will have as a solid. This lowers the density as the water expands and the cold water will float. This expansion on cooling is unusual and is known as **anomalous expansion**. The deeper, warmer water is unaffected as further cooling produces floating solid ice. This floating layer of cold water and ice prevents the water below from freezing and allows most of the water in the lake to remain unfrozen even if the air above is very cold.

If cooling water became denser until it froze, then ice would sink, more water would be exposed to the cold air, this would also sink and eventually the whole lake would freeze, killing most aquatic life.

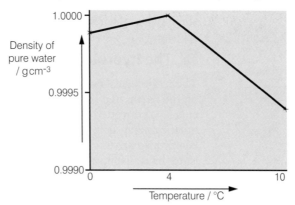

Fig. 8.1 *The anomalous expansion of water*

AQA Examiner's tip

The concepts of density and buoyancy can be difficult to apply to real-life situations. Make sure you understand how the change in density as water cools causes ice to float and reduce further cooling beneath the ice.

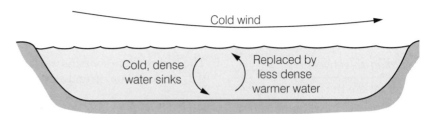

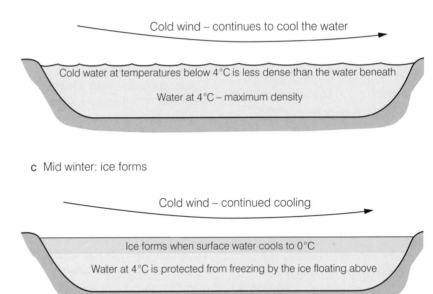

Fig. 8.2 *How ice helps to prevent lakes from freezing*

Solvent properties

Water is an excellent solvent and is called the 'general physiological solvent' because most biological reactions occur with solutes dissolved in water. Plant nutrients can only be absorbed if they are dissolved in water and most materials transported in blood and sap are dissolved in water.

High heat capacity

Water heats up and cools down slowly. This helps to maintain climatic stability by moderating temperature changes.

■ The hydrological cycle

The hydrological cycle is driven by solar power, which warms the water on the Earth, making it evaporate and rise in the atmosphere as the energy is converted to gravitational potential energy. This is converted to kinetic energy as it falls to Earth and flows back towards the sea. Solar energy also drives evaporation from the land and the plants that lose water by transpiration.

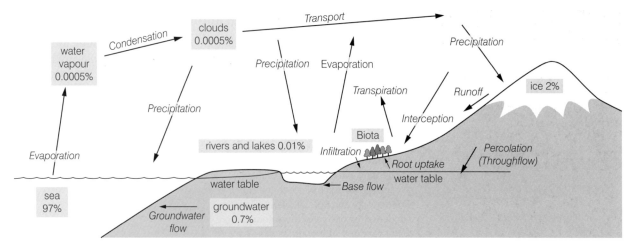

Fig. 8.3 *The hydrological cycle*

The transfer of water between reservoirs

Water is a renewable resource but human actions can alter the rates of water movement in the cycle so that resources become depleted. An understanding of **residence times** can help with the sustainable management of water resources.

Case study

Aquifer depletion

The amount of water in an **aquifer** may be a poor guide to how it can be exploited. Aquifers in the UK are rapidly recharged by rainwater percolating into the ground, so they are reliable sources of water.

Aquifers in some hot, dry countries contain a lot of water, but it has taken a long time to accumulate and a very small amount is added each year. Any exploitation is likely to be unsustainable unless the aquifer is very extensive and is recharged in a distant area with high precipitation.

Key terms

Residence time: the average length of time that a molecule remains in a reservoir.

Aquifer: an underground rock structure from which water is abstracted.

Reservoir: the general name for a storage location for any material, e.g. nitrogen, crude oil, water, iron etc. The term is also used for the stored water retained by a dam.

Transfer rate: the volume of material moved from one reservoir to another.

The average length of time that water remains in a **reservoir** is called the residence time (RT). RT = Vol/average **transfer rate** in or out of the reservoir.

Table 8.1 *Statistics on the hydrological cycle*

Reservoir	Quantity of water /% of total	Volume of water /10^6km³	Residence time
Oceans	97.00	1,370.00	4,000 years
Land ice	2.00	29.00	20–100 years
Groundwater	0.70	60.00	100–10,000 years
Lakes and rivers	0.01	0.125	2 months–100 years
Soil moisture	0.005	0.065	1–2 months
Atmosphere	0.001	0.013	10 days
Living organisms	0.00004	0.006	1 week

■ Hint

You need to understand how the processes in the hydrological cycle move water in the cycle, but it is more important to understand how human activities affect these processes.

■ Key terms

Precipitation: condensation of atmospheric water vapour that reaches the Earth's surface in any liquid or solid form.

Throughflow: the sideways movement of water in the ground.

Interception: precipitation that does not reach the ground because it lands on vegetation.

Infiltration: the process by which surface water enters the ground between the particles of soil or rock.

Percolation: the movement of water between the particles of soil or rock.

Groundwater flow: the movement of water through the pore spaces and fissures in permeable rocks.

River channel discharge: the volume of water flowing past a particular point on a river and is a product of cross sectional area and mean water velocity.

Abstraction: the removal of water from the location where it was found.

The main processes involved in the hydrological cycle

The parts of the hydrological cycle that are on or in the ground are of greatest direct importance to humans. The cycle can be considered as inputs, throughputs or outputs to this part of the cycle.

Table 8.2 *Processes in the hydrological cycle*

Process	Human impacts
INPUTS **Precipitation** includes water vapour condensing and falling to Earth mainly as rain, but also as hail, sleet and snow. Temperature and humidity control when condensation occurs as the amount of water vapour that the atmosphere can hold drops as the temperature goes down	Any impacts that change temperatures: global climate change, urban heat islands
THROUGHFLOW **Interception** involves the precipitation that evaporates back to the atmosphere within a few days having landed on surfaces such as leaves, leaf litter, rock	Deforestation, afforestation, urbanisation with increased paved areas and buildings
Infiltration is the flow of water through the ground surface into the soil or rock below	Soil compaction, urbanisation with increased paved areas and buildings
Percolation is the flow of water through the spaces in soil and rock	Soil compaction
Groundwater flow is the movement of water through the pore spaces in rock.	Groundwater abstraction, artificial aquifer recharge
Runoff is the flow of water over the ground surface	Actions that reduce infiltration or interception
OUTPUTS Evaporation is the change of liquid water to water vapour as hydrogen bonds are broken	Reservoirs, increased temperatures caused by global climate change
Transpiration is the loss of water vapour from plants to the atmosphere through stomata in their leaves	Vegetation changes
River channel discharge is the volume of water carried by a river	**Abstraction** of water, flood drainage

The reservoir provides many recreational and educational opportunities. Some, such as motor boating, may be banned because of the pollution risk to the water. Conflicts may be resolved by Time Zoning or Space Zoning (see Chapter 6).

Case study

Conflicts of interest

The Syr Darya River flows through four countries in central Asia. The Toktogul reservoir in Krygystan is used to generate electricity and the government would like to reduce summer flow to increase winter electricity generation. The countries downstream use the water for crop irrigation and want more water to be released during the summer. Similar disputes have occurred between Turkey and Iraq, the USA and Mexico, Israel and its neighbours, and the countries that share the River Danube.

Summary questions

1 Explain why the properties of water are important for the survival of life on Earth. *(6 marks)*

2 Explain how the quality of water affects the uses to which it is put. *(8 marks)*

3 Explain how the quality of water affects the treatment processes needed to produce potable water. *(10 marks)*

4 Outline how land use conflicts may be caused by the abstraction of water from:
   - aquifers
   - reservoirs. *(4 marks each)*

Sources of water

The availability of water has a major effect on many aspects of life because it influences where people can live, what they can eat, which industries can be developed and the quality of life. The quantity, reliability, purity and location of water all affect the usefulness of a water supply, so the search for water with suitable qualities is an important part of the development of society.

Rivers

River water is often the most convenient source of water. Being long, rivers are accessible over a large land area. The short residence time of water in rivers means they self-clean quite quickly, so natural contaminants are rarely a serious problem. Human pollutants were not a problem until population density rose to the point where the quantity of waste from a community upstream was so great that it could not break down before it reached the next community.

The main features that affect the usefulness of a river are:

- total annual water flow (river discharge)
- flow fluctuations
- level of natural contaminants
- pollutants from human activities.

Fig. 8.7 *A dam built where the valley was narrow*

Activity

Place a piece of tracing paper or an OHP transparency over an Ordnance Survey map that has a reservoir. Trace the reservoir and the catchment area. Add details of important land uses.

Use your tracing to produce an annotated map to show why the site was suitable for a reservoir.

Reservoirs

Reservoir location

Reservoirs allow the storage of water from times when there is a surplus of water until a time when there is a shortage. But even if there is a suitable water supply, many other factors must be considered when choosing a site for a dam and reservoir as follows:

Topography

The main cost of developing a reservoir is in dam construction while the income of the reservoir comes from the amount of water it can hold and then supply.

The ideal **topography** would involve a narrow exit to a large deep basin so a relatively small dam could hold back a huge volume of water.

Geology

The rock beneath the reservoir must be impermeable so the water cannot percolate into the rock and be lost. The rock should also be strong enough to support the weight of the dam and reservoir, without faults or seismic activity that could trigger an earthquake and cause the dam to collapse.

Catchment area

This is the area of land over which falling rain will flow or will flow through the ground and then into a river. So the ability of a reservoir to provide water is controlled partially by the reservoir site itself but also by the surrounding area that collects water for it. Even if it has not rained recently, there may be a lot of water from previous rain that is on its way to the reservoir.

Water supply

Ideally, the rainfall or river inflow should be regular with a large volume. The climate should not be too hot or dry, which would cause excessive evaporation losses.

Existing land use

The use of the land that is to be flooded cannot be so important that it must not be lost. An analysis of the balance of losses and benefits must be considered. In the UK large urban areas and important wildlife conservation areas would probably be protected, while agricultural land would be less valued. Other countries may assess their priorities differently.

Pollution risk

The land uses in the catchment area should not pose a serious pollution risk to the water. The main threats are toxic pollutants from industry and agricultural pesticides. Pollutants such as sewage and manure are biodegradable and break down relatively quickly. They are unlikely to become as concentrated as they could in a river where the volume is smaller.

If the flooded area was covered with forest or a lot of dead vegetation was washed in, then it may decay anaerobically. This would release methane and add to global climate change.

Sedimentation

Soil erosion in the catchment area can make the inflow river very turbid, resulting in sedimentation in the reservoir. This gradually reduces the volume of water that the reservoir can hold.

Infrastructure

Building the dam, treating the water and transporting it to the area of demand require workers, building materials, access routes and machinery. A convenient site near the area of demand may be chosen rather than a site that would supply more water but that is isolated and difficult to reach.

Case study

Water demand in Devon and Cornwall

The best hydrological sites for reservoirs are often those that have greatest value for scenery and wildlife. This makes the decision-making process difficult.

The demand for water in Devon and Cornwall is growing as population increases. Per capita water use is rising as affluence increases and more water-consuming appliances are used.

Demand is highest in the summer when tourists increase the population but rainfall is lowest and evaporation highest. There are no large rivers and few large aquifers. The extra water supplies are most easily met by the storage of surplus winter rain in reservoirs. However, the best reservoir sites are in the upland areas that have greatest scenic beauty and are designated as National Parks or Areas of Outstanding Natural Beauty.

Choosing possible new reservoir sites involves a compromise of the best sites that do not involve unacceptable landscape impact.

Estuarine barrages

A dam across an estuary can be used to create a freshwater lake that can be used for water supplies. An example is the IJsselmeer in the Netherlands (this was also part of a large land reclamation scheme). Few have actually been built because of the pollution risk and interference with navigation.

The environmental effects of reservoirs

Building a reservoir alters the environment of the reservoir site itself and the surrounding area, in the following ways:

Habitat change

Flooding the reservoir obviously destroys the previous habitats but it also creates new and possibly valuable ones. Wetlands are uncommon habitats in most regions so the reservoir may be more valuable than what has been lost.

The dam and reservoir act as a barrier to wildlife such as salmon and sturgeon that migrate along the river. Species that live in the river will have population booms and collapses. Free movement along the river is an important part of recolonising areas that have become vacant in bad years. The dam may prevent this.

Changes in river flow

The reservoir may be used to provide water in two different ways. First, water may be abstracted from the reservoir and piped to the consumers. The flow regime downstream would be the same although the volume would be reduced. Secondly, water may be used to regulate the river flow,

> ### Hint
>
> Do not confuse estuary barrages that produce a freshwater lake with those used to generate electricity where water flows in and out with the tidal flow.

holding water back during times of surplus to ensure adequate river flow in times of shortage so that water can be abstracted from the river further downstream. This reduces the risk of flooding downstream but also reduces periods of lower flow that are important for some species such as river turtles that lay their eggs in sandbanks. Periods of rapid flow are also important to wash away sediments from gravel riverbeds in which salmon and trout lay their eggs. The changes in flow fluctuations can also change river erosion and sedimentation and therefore the development of meanders.

Sedimentation

The sediments that settle in the reservoir will no longer be carried further downstream. In the past they may have been important to fertilise the floodplain downstream during times of flood. They may also have been important in helping to build up riverbanks and coastlines and counteract erosion.

Reservoir microclimate

The large body of water in the reservoir may change the local climate. The high heat capacity of water helps to reduce temperature fluctuations. Therefore it will be warmer in winter and cooler in summer. Water provides less friction than land, so wind speeds will be higher. Greater evaporation from the reservoir surface may increase humidity, cloud cover and precipitation downwind of the reservoir.

Aquifers

An aquifer is a layer of rock that holds water, which is exploited as a resource. To be suitable for exploitation it must have certain features.

The main features of aquifers

Porosity

Porosity is a measure of the proportion of the rock's volume that is space and could therefore hold water. Chalk, limestone and sandstone are porous rocks that often form aquifers.

Permeability

Permeability is a measure of the ease with which fluids may flow through a rock because of the interconnections between the spaces. Some materials such as clay have high porosity but the pores are too small for water to flow through easily.

> ### Key terms
>
> **Porosity:** a measure of the percentage of the volume of a rock that is space.
>
> **Permeability:** a measure of the rate at which a fluid, such as water, can flow through rock.

Suitable geological structures

The rock below the water-bearing rock must be impermeable to prevent the escape of the water. Granite and clay are suitable impermeable materials.

Some of the rock above must be permeable to allow recharge of the aquifer with water from above. Some aquifers are very large and the recharge area may be a long way from the area of abstraction. The water may be abstracted using a well, a borehole or it may come to the surface naturally in springs.

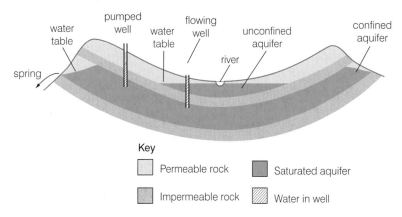

Fig. 8.8 *The geological structures associated with aquifers*

Mineral resources

Mineral resources include the rocks and fossil fuels that are removed from the crust to be used, after processing if necessary.

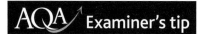

Table 9.1 *The most important mineral resources*

Resource	Most important uses	Important properties
Fossil fuels		
Coal	Fuel, especially for electricity generation. Coke for iron smelting	High energy content
Crude oil	Fuels, especially liquid fuels for vehicles Petrochemicals and plastics	High energy content Carbon forms very complex molecules
Natural gas	Fuel, especially for heating and electricity generation	High energy content
Metals		
Iron	Construction: buildings, ships, road vehicles, appliance cases, e.g. cookers	Malleable, can be moulded and rolled Strong but high density Rusts Forms alloys, e.g. stainless steel
Aluminium	Packaging Vehicles Construction – window frames etc.	Malleable Strong with low density Good conductor of heat and electricity Corrosion-resistant
Copper	Electric cables Water pipes	Malleable Very good electrical conductor Corrosion-resistant
Chromium	Stainless steel	Forms alloys – makes steel harder and corrosion-resistant
Non-metal minerals		
Sand	Builders' mortar Glass	Fine-grained inert filler Transparent in thin sheets
Gravel	Concrete	Coarse-grained filler
Clay	Bricks, roof tiles, pottery	Waterproof when baked, easily moulded
China clay	Filler in paper, pottery, plastic items, paints, cosmetics, pharmaceuticals	Very fine-grained, inert, non-toxic filler
Limestone	Building blocks, crushed for aggregate for road surface chippings and railway track ballast Cement	Hard, easily cut and crushed, quite resistant to wear Baked lime (with clay and gypsum) hardens when mixed with water
Granite	Blocks for buildings and road kerbs, kitchen worktops, floor tiles, crushed for road surface chippings	Hard and resistant to wear
Slate	Roof slates	Waterproof, splits to form thin, flat sheets

The geological origins of economically important minerals

If all the minerals in the crust were evenly mixed then none of them would be sufficiently concentrated to allow exploitation. Geological processes have provided local concentrations that can be exploited. They may still be present in very small amounts and therefore be difficult to find, but an understanding of the processes that produced them can help as it will show which other geological formations would be expected in the surrounding areas.

Igneous processes

The plates that make up the crust are moved slowly by convection currents in the mantle below. The friction, heat and pressures produced can cause molten rock called magma to be forced up towards the surface. It may reach the surface as an **igneous** extrusion where it will cool rapidly and form a fine-grained rock such as basalt.

Fig. 9.2 *A basalt igneous extrusion on the island of Staffa, Scotland*

Some magma may fail to reach the surface, forming a large molten mass of intrusive rock. This is insulated by the surrounding rock so it cools slowly, producing larger crystals and giving more time for minerals to be separated into localised concentrations that can be exploited.

Granite was produced in these deep deposits called **batholiths**. The rocks around the batholith may be lifted and deformed so cracks or fissures form. These may allow hot solutions containing dissolved minerals to escape towards the surface. As the solutions cool down, the minerals are deposited in a predictable order according to their solubility. This process takes solutions of mixed minerals and separates them into concentrated deposits that can be exploited more easily. Most metals, for example iron, tin, copper and lead, are found as **hydrothermal** deposits.

> ### Key terms
>
> **Igneous:** rocks or processes involving molten rock.
>
> **Batholith:** a large underground mass of solidified molten magma.
>
> **Hydrothermal:** processes or deposits associated with hot water.

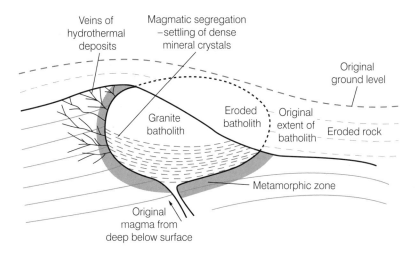

Fig. 9.3 *A granite batholith and associated hydrothermal deposition*

Fig. 9.4 *The granite outcrops on Dartmoor are part of a very large batholith*

Sedimentary processes

Sedimentary processes occur at the surface of the lithosphere. A rock that already exists is broken down by weathering into rock particles (**regolith**) and dissolved minerals in solution (**solutes**). These are carried away, separated and may be redeposited elsewhere as rocks with quite different compositions and properties.

- **Alluvial/placer deposits** have been carried by flowing water but are deposited when the water slows down. Different minerals are deposited in different parts of the river, with the densest ones settling out first. Tin ore, gravel, sand and clay are all deposited in this way. Mineral solutions may later be deposited in the spaces between the rock particles, sticking the particles together. This is what happens to sand when it forms sandstone.

- Evaporites are formed when water evaporates from mineral-rich solutions. Minerals crystallise as the solutions become saturated, each mineral reaching saturation at a different time so they become separated in layers. For example, gypsum (calcium sulfate – building plaster) is deposited before halite (sodium chloride – cooking salt).

- Biological deposits are produced from dead animals and plants. Fossil fuels were produced by dead organisms partially decaying in anaerobic conditions. Chalk and many limestones are formed from the shells of dead marine organisms.

- Chemical precipitates may be deposited such as the manganese nodules found on some deep ocean beds.

Metamorphic processes

Existing rock that is exposed to extreme heat and pressure from nearby igneous activity may change its form without melting, in a **metamorphic** process. Slate was formed from **sedimentary** shale, often made largely from clay.

Key terms

Regolith: the solid rock particles left after weathering.

Solute: a dissolved substance.

Alluvial: Materials such as soil or weathered rock particles deposited by a river or other flowing water.

Placer deposits: deposits of dense minerals carried by water, e.g. tin and gold.

Metamorphic: rocks changed by intense heat and pressure, but without fully melting.

Sedimentary: materials or processes that involve material being carried by air or water then deposited.

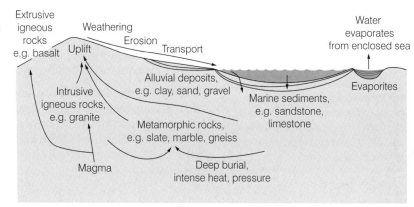

Fig. 9.5 *The rock cycle and the formation of mineral resources*

Hint

Make sure you understand the difference between resources and reserves. The words are both used rather vaguely in everyday conversation but they have precise meanings.

Key terms

Reserve: the proportion of a resource that can be economically exploited with existing technology.

Did you know?

The mining industry has to plan well into the future because it takes a long time to develop a site for mining, so exploration must be carried out to find suitable sites for future mines. Those deposits where enough is known to calculate the amount that can be extracted economically are called 'proven reserves'. Deposits that have been found but where the amount that could be extracted has not been established are called 'probable reserves'. 'Speculative reserves' are where geological research has not found deposits that could be economically exploited but the surrounding geological structures make it likely that they do exist.

Hint

The definitions of different types of reserves are less important than an understanding of the fact that the amount of a material that can be exploited is not fixed. It depends on geology, technological abilities and the choices that are made over environmental issues.

Resources, reserves and exploitation

Minerals are non-renewable resources because the amounts that exist are finite although most are very abundant. Economically recoverable resources account for a tiny proportion of the total that exists.

The main limitations on mineral availability are the locations, chemical form and purity of the deposits, and the availability of technologies to exploit them.

Their exploitation is economically important but can cause environmental damage.

Resources and reserves

Resources

The term 'resources' is used in general terms to mean the materials that can be exploited. If the term 'resource' is used when referring to a particular material then it includes all of that material that is theoretically available for exploitation. This includes deposits that can be exploited as well as those that cannot be exploited now but could be with realistic increases in prices or improvements in technology. Resources that are deep, low grade, in a difficult chemical form or in locations that are currently protected could all be usable in the future.

Reserves

Reserves include that portion of the resource that can be exploited now, economically, using existing technology.

The size of a resource is finite but the quantity included in the reserves can change. If there is an increase in market price or if new extraction technologies are developed then the reserves will increase. If market prices drop then reserves may decrease as deposits that still exist become uneconomic.

Factors affecting mine viability

Mineral deposits are likely to be mined if it will be sufficiently profitable and there are no prohibitive environmental, political or social problems.

The profitability of mining is simply the difference between the income from selling the processed mineral and the total costs of extracting and processing it. The people that wish to profit from mining may not be those that are most concerned about the problems created.

Land conflicts

Deciding where to mine minerals is different from most other land-use choices. If you want to build a reservoir, town, road or airport or need to decide where to farm or plant forests then there are choices. But the only place you can mine minerals is where they are found.

If competing land uses are considered to be more important or valuable than mining then the deposit may not be exploited. Deposits below urban areas are not usually mined as the cost of moving the population would be too high. Areas with high landscape or wildlife conservation value may also be protected, although this is not considered to be important in all countries.

Extraction costs

Overburden

The rock above the mineral that must be removed is called the **overburden**. If the overburden is very hard it is expensive to remove as it may need to be blasted.

Depth

Mining costs rise rapidly as depth increases. If the depth is doubled then the cost much more than doubles. The sides of the mine cannot be vertical because of the risk of collapse, so the amount of rock that must be removed to reach the mineral rises rapidly as depth increases. If the overburden is loose then the gradient must be more gentle, which increases costs even more.

Form of the mineral deposit

Mining costs will increase if the mineral is found in thin layers or if it is dispersed in an irregularly shaped deposit. Both of these problems would increase the size of the mine void that would need to be excavated.

Hydrology

As depth increases, the amount of water that flows into the mine also rises. Pumping costs can be high.

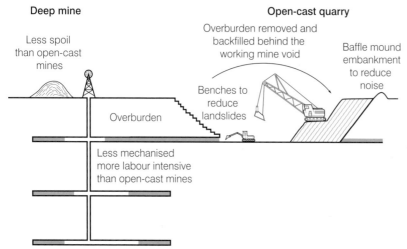

Fig. 9.6 *Features of the main types of mine*

Fig. 9.7 *Mechanisation allows the rapid removal of large amounts of material*

Processing costs

The chemical form of the mineral

The cost of extracting a metal depends upon the other elements with which it is combined. The more energy that is needed to break the bonds the more expensive it will be to extract the metal.

Purity

The financial cost of extracting metal from rock increases rapidly as the purity of the ore decreases. To produce one tonne of metal, a greater quantity of rock must be extracted and more energy is required for the chemical separation of the metal from its ore.

The cut-off ore grade is the lowest ore purity that can be exploited economically. Deposits that are below the **cut-off ore grade** are not included in the reserves so, as prices fluctuate, the quantity of mineral in the reserves also changes although the amount of material that exists does not.

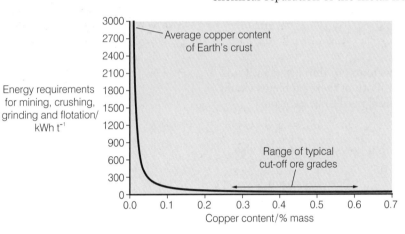

Fig. 9.8 *The effect of ore purity on the energy cost of purifying metals*

■ Key terms

Cut-off ore grade: the lowest purity of ore that can be exploited economically.

Transport costs

The distance to market, the ease of bulk transport and the presence of a suitable existing transport infrastructure all affect transport costs.

Transporting minerals longer distances increases costs, but the unit costs go down if bulk transport by rail or large ship is possible. If there is an existing transport system then the set-up costs are lower and processing the mineral before it is transported reduces the bulk that needs to be moved.

Market economics

The market demand and sale value of the minerals control the economic viability of exploiting a particular mineral deposit. The market price is controlled by the demand for the mineral and how much is produced by mines. Supplies rise and fall relatively slowly as mines are developed, but demand can rise and fall quickly. This mismatch can cause prices to fluctuate widely. This uncertainty makes the prediction of future markets very important.

Exploiting deposits in regions that already have mines is easier because there will be easy access to the existing infrastructure for transport, energy, equipment supplies and a trained workforce.

Mineral extraction

A brief history of mining methods

The first minerals to be exploited such as flint were found on the surface. When these were exhausted shallow pits were excavated. As these got deeper, they became vertical shafts, with horizontal tunnels where the minerals were found in horizontal seams. Human labour provided the energy.

Very deep shafts were only practical when pumping water out of the mines with steam engines became possible. Equipment was developed to help cutting or blasting the tunnels and minerals, and to carry materials out of the mine.

Labour costs gradually increased and machinery became larger so it became more economic to mine mechanically, using human labour mainly to drive the machines. Almost all modern mines use open-cast methods, also called open-pit mining or quarrying. Deep mining with vertical shafts and horizontal tunnels can be used for deeper mineral deposits but it is much more expensive and cannot use huge machines to extract massive amounts. So, it is rarely used, except for high value minerals such as gold and diamonds.

Fig. 9.9 *The winding gear of a deep mine could not extract large quantities of minerals*

The environmental impacts of mineral exploitation

Exploration, extraction and processing all cause significant environmental damage. A range of methods may be used to reduce these although they all cost money, so the protection of the environment is balanced with market profitability. Profits are greatest if less money is spent on environmental protection, but planning authorities in many countries only grant permission to mine if the environmental damage is minimised during the active mining period and in site restoration afterwards.

Exploration

Marine seismic surveys cause very loud vibrations and can disturb whales. Exploration on land can involve land clearance and vegetation loss.

Land impacts

Extraction may cause conflicts with other land uses.

Land take

The land area that is required is larger than the area of the mine void (hole). Land is needed for associated buildings, access routes, overburden dumping and possibly a buffer zone between the mine and neighbouring areas.

Habitat loss

The loss of species where the mineral is to be extracted is unavoidable.

Removing the wildlife by capturing the animals and transplanting the plants to move them to unthreatened habitats has been attempted but is rarely completely successful. Such areas are usually either unsuitable or are already fully populated.

Habitat restoration when mining has ended is often carried out, or new habitats may be created, such as a wetland nature reserve in the mine void. This is a requirement of planning permission in many countries, including the UK. The newly created habitat may have greater wildlife value than the habitat that was present before mining started.

Loss of amenity

Mining may cause aesthetic problems for local communities. This may be reduced by landscaping and tree planting. If the mine is turned into a community resource when mining has ended then the long-term amenity value may be increased.

Pollution

Dust

Dust is raised into the atmosphere by blasting and vehicle movements. The dust can be removed with water sprays.

Noise

Mine vehicles and rock blasting are the two main sources of noise. Embankments or 'baffle mounds' built around the mine help absorb and deflect noise.

Blasting is not continuous but it can be disturbing if it is unexpected. This can be reduced by blasting at set times so the noise is predictable.

Turbid drainage water

Mining produces small particles that may be carried away in drainage water that would increase the **turbidity** of the river into which it flows. High water turbidity blocks sunlight from reaching aquatic plants and chokes filter-feeding animals.

The turbidity of drainage water can be reduced by keeping it in a holding lagoon. Because the water is still, the suspended solids settle out, leaving the clear water to flow into the river.

Toxic leachate

A toxic metal that is normally immobile in rocks may be oxidised in a spoil heap and become soluble. Drainage water can then carry it into a nearby river as toxic **leachate** where it may kill aquatic life.

In hot areas the drainage water can be collected and evaporated. The metal solution will become more and more concentrated until it may be possible to extract the metal economically.

Many toxic metals are more soluble under acidic conditions. Passing mine drainage water through a filter bed of crushed limestone can immobilise the metal and prevent it flowing into the river.

Spoil disposal

Spoil heaps are often loosely compacted and instability can cause landslides or erosion.

Drainage of rainwater by pipes in the base of the spoil heap prevents it becoming waterlogged, which would make it more likely to slip. Compaction of the spoil and landscaping to avoid steep slopes will also reduce instability.

Flooding

Flooding can be caused by drainage water if it is pumped out rapidly or if it is stored behind a poorly constructed dam that may collapse. The risk can be reduced by containment in lagoons behind well-constructed dams with carefully timed releases.

Water that flows into an abandoned mine may overflow into a river carrying dissolved toxic metals and acids.

Subsidence

This is caused by poor spoil compaction or undermining of sensitive surface land uses by deep mines. It can be reduced by compaction of spoil and by leaving support pillars in deep mines.

Key terms

Turbidity: a measure of the cloudiness of water caused by suspended solid particles.

Leachate: liquids and dissolved materials such as fertilisers washed through the ground, usually downwards.

Energy resources

Major characteristics of energy resources

Renewable or non-renewable?

Non-renewable energy resources are reformed by natural processes very slowly. They are usually formed by geological processes and the rate of formation is so slow that we cannot wait for new resources to be formed.

Non-renewable resources include nuclear fuels and the fossil fuels: coal, crude oil, natural gas, oil shales and tar sands. These are 'finite' resources because the amount that is available for use is fixed, so as they are used the remaining reserves are **depleted**.

Renewable energy resources are reformed by natural processes fast enough for new supplies to become available within a human lifespan. Some are reformed instantly, such as sunlight, while others, such as wood, take longer.

Some of these renewable resources cannot be depleted by using them. Using solar power, wind power or tidal power will not make it less sunny, less windy or stop the tides from flowing. But other renewable energy resources can be depleted if they are used unwisely. If trees are felled faster than they are growing then supplies will be depleted.

Energy density

Energy density is a measure of the amount of energy that can be harnessed from a given quantity of resource. It is easy to quantify as Joules per kilogram of fuel, such as uranium, coal, oil, wood or straw. It is more difficult to produce comparative figures for resources such as wind power and solar power, but energy harnessed per kilogram of equipment can be used. In general the higher the energy density the more useful the resource is.

Table 10.1 *The energy density of some energy resources*

Energy density category	Energy resources
Very high	Nuclear fusion
	Nuclear fission
High	Hydrogen
	Fossil fuels
Medium	Wood
	HEP
Low	Solar power

Hint

You do not need to know the technical details of the individual energy resources described here unless they help you in understanding the environmental issues that affect their availability, usefulness and environmental impacts.

Key terms

Non-renewable resource: a resource that is renewed so slowly that the amount available for use is effectively finite.

Deplete: to reduce the amount of a resource that is available.

Energy density: the amount of energy in a particular mass of fuel or that can be harnessed by a particular mass of equipment.

Fig. 10.9 *Straw has a low energy density so long-distance transport is not worthwhile*

If a fuel has a high energy density then smaller quantities will be required. This can reduce the cost of fuel transport and make users of energy less dependent on fuel deliveries.

A low energy density fuel is generally less useful, but it should not be rejected without further consideration. Wind power has a low energy density, but the electricity generated can be used to electrolyse water and produce hydrogen, which has a high energy density and can be used to drive vehicles.

Intermittency

If an energy resource is not available all the time then it cannot be relied upon and other resources must be available to meet the total demand. Many renewable energy resources are **intermittent**.

Reliability

Unreliable resources are those where the amount of energy that will be available cannot be predicted. This is not the same as intermittency, as some intermittent resources, such as tidal power, are intermittent but very **reliable**. Other resources, such as wind power, are both intermittent and unreliable.

Ease of storage

The supply of energy and demand for it are not constant or perfectly matched. It may therefore be necessary to store energy from when it is available until the time it is needed.

Storage may be needed to make intermittent supplies more useful, or to produce energy at a constant rate so that they can be 'peak shaved': the surplus that is not needed can be stored until supplies cannot meet demand.

Available resource

The amount of energy that is available obviously affects its potential contribution to energy supplies. It is not always easy to estimate the proportion of the available resource that can realistically be harnessed.

Geographical and locational factors

A range of locational factors affect the ability to exploit energy resources. For example, resources that need to be extracted, such as fossil fuels and uranium ore, can only be exploited where they are found in favourable deposits. (See Chapter 9, Factors affecting mine viability.)

Resources that harness natural processes may rely on factors such as the climate and topography.

Level of technological development

It takes a long time to develop any technology until it is as efficient and reliable as possible. For more complex technologies this can be a slow process as it may take a long time to plan, build and use a piece of equipment before it can be redesigned to make it better for the next generation of use.

Environmental impact

All energy resources cause environmental damage in a variety of different ways. Pollution caused during the use and extraction of fuels is often very obvious. There is also damage caused during the manufacture of the equipment required to extract the fuel.

Some of the environmental impacts are over-emphasised. The **aesthetic** problems caused by installing equipment such as wind farms upset many people, often the same people who are using the energy, but it has no impact on the ability of the planet to sustain life.

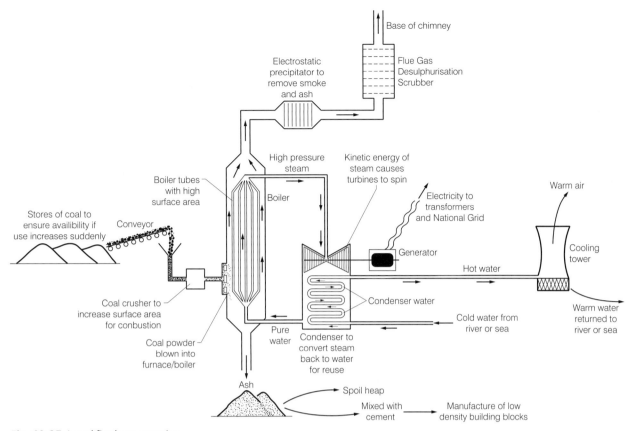

Fig. 10.25 *A coal fired power station*

In conventional power stations, heat is used to boil water. The movement of the high pressure expanding steam is used to spin turbines which spin the generators.

Electricity from light energy

Photons of light that strike the surface of the photovoltaic cell displace electrons, which make the surface layer negatively charged. These free electrons can flow along a conductor to the lower layer powering an electrical appliance in the process.

Electricity from chemical energy

Chemical energy can be converted to electricity in reactions where one chemical accepts electrons and another releases them. The electrons flowing from one to the other can be used to power electrical equipment. Batteries (electrochemical cells) store chemical energy that can be converted to electricity, but as the chemical reactions occur, the battery loses its ability to produce electricity.

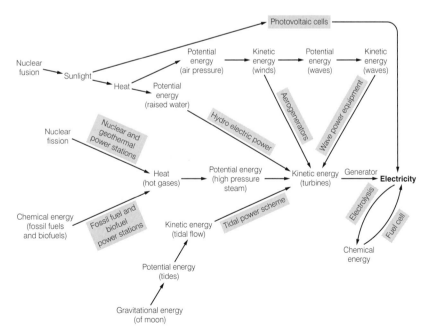

Fig. 10.26 *The conversion of primary fuels into electricity*

A **fuel cell** is similar to a battery but it will not go flat as long as there is a continual fresh supply of the chemicals that are involved in the chemical reactions.

The simplest fuel cell uses hydrogen and oxygen. The hydrogen gives up electrons that are accepted by the oxygen. The hydrogen and oxygen ions produced combine to produce water. Other fuel cells use hydrocarbons and alcohols.

Advantages of using electricity:

■ It causes no pollution at the point of use.

■ It is easy to convert electricity to other forms of energy such as heat, light, movement and sound, so it has many applications.

■ It is easy to transport along conducting cables.

Disadvantages of using electricity:

■ The efficiency of converting the primary fuel into electricity is often very low.

■ There are no large-scale methods of storing electricity.

■ The original method of generating the electricity may cause pollution, such as burning fossil fuels.

Hydrogen

Hydrogen is a reactive gas that can be produced easily by the electrolysis of water. When it releases its chemical energy it recombines with oxygen to form water again.

The electrolysis can be driven by electricity from surplus primary energy sources that could not be stored. This overcomes the major disadvantage of unreliable or intermittent energy supplies by converting them into a form of energy that can be stored.

The chemical energy of stored hydrogen may be released in two main ways:

■ Combustion to produce heat – this could be used for space or process heat or it could be used to generate electricity by boiling water to drive steam turbines.

■ Fuel cells – these allow oxygen and hydrogen to combine, producing water and releasing energy, but the process is electrochemical rather than combustion.

The electrolysis of water to produce hydrogen provides a method by which surplus electricity from intermittent primary energy sources may be stored and therefore be available on demand. This could allow resources such as solar, wind, wave and tidal powers to provide significant proportions of the total energy demand.

■ Energy storage

The storage of energy helps solve several problems:

■ to build up a sufficient quantity for transportation

■ to allow energy production rates to be kept constant for the most economic use of equipment and workers, e.g. coal mines, oil and gas wells

■ to store energy from intermittent supplies so that an energy surplus can be stored until it is needed, e.g. wind and solar power. Storing a surplus to meet a later shortage is called **peak shaving**.

Pumped storage hydro electric power

While fossil and nuclear fuels can be stored relatively easily, electricity cannot be stored on a large scale.

Surplus electricity may be available for several reasons:

- base-load power stations may be generating electricity at night when it is not needed, but it is uneconomic to turn the power stations off as they will be needed again soon

- a reduction in demand after meals or TV breaks may occur more rapidly than power station output can be reduced.

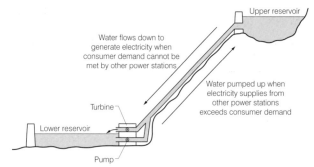

Fig. 10.27 *A pumped storage HEP station*

If this electricity is not used then it would simply be lost as heat in the National Grid.

The electricity is used to pump water uphill, thus storing the energy as potential gravitational energy.

The PE (potential energy) can be converted to KE (kinetic energy) by allowing the water to flow down to the lower reservoir generating electricity as it turns turbines. This can be done when there is a sudden increase in demand for electricity such as at mealtimes and during breaks in popular TV programmes. These power stations can respond to changes in demand much more rapidly than fossil fuel or nuclear stations.

Hydrogen

Hydrogen produced using surplus electricity can be stored for later use. See The hydrogen economy, on page 160.

Chemical energy

Batteries and fuel cells can be used to store chemical energy that can be converted to electricity. Rechargeable batteries involve chemical reactions where the original chemicals can be reformed using electricity so they can be reused. Fuel cells require a supply of new fuel and oxidant chemicals for continued use. These chemicals are often made using surplus electricity, e.g. by the electrolysis of water.

■ The environmental impacts of energy use

All energy resources have environmental impacts although they vary in their importance and how obvious they are in importance.

The main activities associated with energy exploitation that have an impact on the environment are:

- fuel extraction
- site development and operation
- material transport
- waste disposal.

It is also worth considering some specific impacts associated with particular resources.

> ### Hint
>
> You do not need to know every impact of every resource but you must be able to illustrate and discuss the major issues using suitable examples.

Fig. 10.28 *A natural gas pipeline under construction*

Fig. 10.29 *A waste flask transporting used nuclear fuel*

The environmental impacts of fossil fuels

Deep mining for coal causes less habitat damage, but surface subsidence above the mine can be a problem.

Oil spills from oil rigs can cause pollution. Oil-based drilling mud used to lubricate the drill pipes can cause pollution in groundwater, rivers and the sea.

Surplus gas on oil rigs may be burnt or 'flared' to reduce the risk of explosions. This causes atmospheric pollution.

The construction of oil and gas pipelines can cause habitat damage over a large total area, although careful soil replacement and replanting can reduce this.

Table 10.4 *Environmental problems associated with open-cast mining*

Problem	Solution or remediation method
Aesthetic problems	Landscaped embankments Landscape the area after mine closure, plant trees and grass
Dust from mining and vehicle movements	Spray water during dry weather
Noise from vehicles	Baffle mounds (embankments) around mine to absorb/deflect noise
Noise from blasting	Limit blasting to daytime periods
Turbid drainage water	Sedimentation lagoons
Derelict site left on mine closure	Restoration for appropriate use – agriculture, conservation, recreation Ground instability makes house construction difficult

The environmental impacts of nuclear power

Wastes that contain radioactive materials come from many sources. The most important ones are:

■ radioactive wastes from fuel manufacture

■ waste fuel after use in the reactor

■ anything that has become contaminated by contact with fuel, e.g. protective clothing, tools and process equipment

■ materials that have been exposed to neutrons from the chain reaction and developed activation products (atoms that become radioactive following neutron bombardment), e.g. fuel rod cases and reactor components.

Radioactive wastes are categorised according to the level of radiation they emit and whether they generate heat.

As long as the waste is contained in sealed containers with adequate absorbing materials or space so the radiation released does not affect the workers or public, then waste storage is safe. The most radioactive isotopes have relatively short **half-lives** so become less hazardous within a few decades. The remaining radioactive materials are not a major hazard but must still be stored safely.

The 'hydrogen economy' is the name given to the proposal that hydrogen storage could be used to allow society to be run on abundant, convenient energy from renewable sources.

The available energy sources would be used to satisfy current demand, but any surplus energy would be stored by producing hydrogen. This hydrogen would be stored until there was a shortfall in supplies caused by unreliable or intermittent supplies or by an increase in demand.

The hydrogen could be used directly by consumers in vehicles or for heating, or it could be converted into electricity in steam-turbine power stations or in fuel cells.

Although a hydrogen economy may be possible, it would not be necessary for hydrogen to be the main energy source that was actually used by consumers. Primary energy would be used when it was available and there was a demand for it. Primary resources that could be stored, such as biofuels, would be stored if there was no current demand.

This approach would reduce the energy losses that would have occurred during energy conversions if all the primary energy had been used to make hydrogen.

Hydrogen has a high energy density and, unlike many renewable energy resources, could replace fossil fuels for many uses, including vehicle fuels.

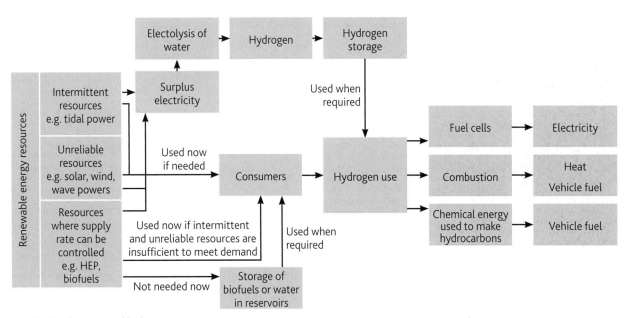

Fig. 10.36 *The proposed hydrogen economy*

Case study

Hydrogen power in Iceland

Iceland is the first country to plan for a hydrogen economy. It is planned that hydrogen will be used to power road vehicles as well as its fishing fleet, which is the country's biggest export income earner and accounts for most fossil fuel use.

Activity

Research the latest developments in the production of hydrogen and its use in fuel cells. Assess how improvements in the efficiency of energy conversions affect the viability of using hydrogen.

Energy conservation

Energy conservation involves reducing the amount of energy needed by increasing the efficiency of energy use and avoiding the wastage of unnecessary use.

■ Efficiency relates to the proportion of the energy being used that actually achieves the desired task. If energy is used efficiently then less needs to be used.

■ Wastage relates to activities that should not be carried out, although this is often a matter of opinion.

A small car may use energy more efficiently than a large car, but driving it just for fun is still wasteful.

Domestic energy conservation

Space heating

Space heating is the biggest domestic use of energy. To achieve a constant internal temperature, the heat inputs must balance the losses. If the losses can be reduced then the inputs can also be reduced.

The rate of heat loss from the house depends on the following factors:

■ the temperature gradient – the temperature difference between inside and out

■ the thermal conductivity or resistance of the materials that form the outside of the building

■ the loss of warm air from inside the building

■ the areas of each type of material, which form the outside of the building

■ the chilling effect caused by wind and rain.

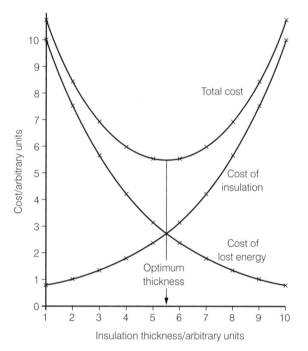

Fig. 10.37 *The effect of insulation thickness on the costs of heat loss*

Energy conservation programmes usually concentrate on the first three of these.

The method that is used to reduce energy loss depends upon the part of the house involved. Through the roof, walls and floors of a house, heat is lost by conduction. Adding a material that has a low thermal conductivity can reduce this. These include polystyrene, mineral wool, shredded paper and wool. Adding extra layers of insulation reduces the rate of heat loss, but each extra layer will save less energy than the previous layers.

Heat through windows is lost by conduction from the warm air inside, through the glass to the cold air outside. A convection current in the room next to the window allows more warm air to come in contact with the glass and lose its heat. Double glazing produces a layer of static air next to the outer pane of glass. The lack of convectional air movement reduces the rate of heat loss.

Because air is a poor conductor of heat, having a larger air gap reduces the rate of heat loss. But if it is too large then a convection current between the window panes can occur and heat loss will increase.

Heat loss will be reduced more if the gap between the panes is evacuated or filled with a gas that is a poorer conductor than air, such as helium or argon.

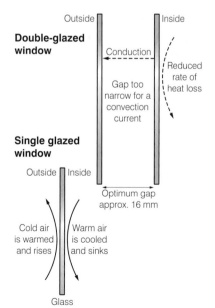

Fig. 10.38 *Single and double glazed windows*

The properties of pollutants

The major pollutant properties

Toxicity

Toxicity is the measure of how poisonous a material is. A toxic substance is poisonous and harms the biological processes that take place within living organisms.

Most toxic substances cause harm by damaging proteins, especially enzymes that control biological reactions. If these reactions cannot take place then the cells affected may malfunction.

Case study

Toxic pollutants

- Carbon monoxide binds to haemoglobin in the blood and prevents it carrying oxygen.
- Lead inhibits the enzymes in nerve cells.
- DDT prevents the normal passage of nerve impulses between nerve cells by inhibiting enzyme action.
- Acids (e.g. from acid rain or acid drainage from mine spoil heaps) change the shape of cell membrane and enzyme proteins so they do not function properly.
- Some toxic pollutants have particular toxic effects and are given specific names. For example toxins that damage the nervous system are called neurotoxins.

Not all pollutants are toxic. Many cause harm indirectly by altering some other environmental factor, which then causes harm.

Case study

Non-toxic pollutants

- Sewage causes **deoxygenation** of lakes and rivers.
- Greenhouse gases cause the Earth to become warmer.
- CFCs cause ozone depletion.

Specificity

Specificity is a measure of the differing toxicity of a substance to different types of organisms.

Different organisms have different sensitivities to toxic substances. So, a particular toxic pollutant may be very harmful to some organisms while less harmful to others. Therefore a particular dose or concentration of a pollutant may kill some organisms while others can survive when a larger dose may have harmed them too.

For example, **pyrethroid** insecticides will kill both insects and mammals, but insects are more sensitive than mammals. Farmers use a dose that is sufficient to kill insects but not high enough to harm mammals such as livestock and humans. Pyrethroids are said to be specific toxins for insects.

Key terms

Toxicity: a measure of how poisonous a substance is, usually caused by enzyme inhibition.

Deoxygenation: the process of oxygen removal.

Specificity: a measure of the differing toxicities of a substance on different organisms. A specific substance is very toxic to some organisms and much less toxic to others.

Pyrethroids: insecticide pesticides, originally extracted from chrysanthemum flowers.

Key terms

Persistence: a measure of the rate at which a material breaks down and therefore the length of time it remains in the environment.

Organochlorines: synthetic organic substances that include chlorine, such as the pesticide DDT.

Degradation: the breakdown of a material.

Biodegradation: the process of breakdown of a material by living organisms, usually bacteria.

Biodegradability: the ease with which a material is broken down by living organisms, usually bacteria.

Persistence

Persistence is a measure of how long it takes for a pollutant to chemically break down.

Many pollutants are compounds with atoms of different elements combined together. A variety of chemical, biological or physical processes may change the compounds so that they cannot cause harm and are therefore not pollutants. Persistence is a measure of the length of time that a pollutant remains in the environment and can therefore cause harm.

The term persistence is usually only applied to substances that will eventually break down, so it is not used for elements that cannot break down or compounds that are so stable they will not break down under normal environmental conditions. These would be described as being non-degradable.

Examples of persistent pollutants include the following:

■ CFCs, which are chemically stable and break down slowly. They may remain in the atmosphere for many decades.

■ DDT is stable, although it does gradually break down. Although it has not been used in the UK for over 30 years, some is still present in wildlife and humans.

■ Heavy metals such as lead and mercury are elements and do not break down at all.

Examples of non-persistent pollutants include the following:

■ Sewage breaks down rapidly as long as the conditions are right for the bacteria that cause decomposition – warm, plenty of oxygen etc.

■ Pyrethroid insecticides break down rapidly compared with **organochlorines** such as DDT.

Biodegradability

Degradation is the process of chemical breakdown.

Biodegradation is the process of being broken down by biological processes, especially the action of decomposers such as bacteria and fungi. The rate of biodegradation or **biodegradability** is affected by environmental factors such as temperature and oxygen availability.

Biodegradable pollutants include sewage, paper and cardboard. Less biodegradable pollutants include most plastics and DDT.

Mobility

Mobility is a measure of how easily a material moves in the environment.

Pollutants that are more mobile are likely to travel longer distances and therefore cause problems over larger areas, although they may also become diluted and less locally harmful.

CFCs are persistent gases which are not very soluble in water and are therefore not removed by rain. So they can travel long distances in the atmosphere.

Because sulfur dioxide is a gas, it can be carried long distances before it causes acid rain. But, because it is washed out of the atmosphere when it dissolves in rain, it usually travels a maximum of a few hundred miles.

Smoke particles are easily washed out of the atmosphere or settle due to gravity. So they rarely travel more than a few miles unless they are

Synergism

Cadmium and zinc can both have harmful effects by enzyme inhibition, but if they are both present then they act synergistically and the health effects are greater.

Solubility and pH

Most heavy metals are more soluble at lower pH. This is useful in treating heavy metal wastes. If the pH is high then the heavy metals will be insoluble and will not be dispersed by water.

Lead

Lead has been used to make many products including car batteries, pipes, solder, paint, anti-knock petrol and flashing for roofs and windows on buildings.

Because of the increasing awareness of the toxicity of lead, some of these uses have been reduced or stopped, while others are not a pollution risk because the lead is not released into the environment.

Pathways leading to the absorption of lead

Industry

Workers may inhale lead dust or absorb it through the skin if it dissolves in sweat.

Dust levels in the air are reduced in lead mines by using water sprays. Workers may also wear dust masks to reduce inhalation.

Water

Lead has been used to make water pipes for thousands of years and more recently as a solder to join pipes. Small amounts of lead dissolve in the water and may be swallowed if the water is drunk.

Water companies may add small amounts of phosphoric acid to the water in areas where old lead pipes are still used. The lead reacts with this to produce an insoluble layer of lead phosphate, which prevents more lead from dissolving.

Most lead pipes have been replaced and no new ones are used. Lead solder is no longer used on drinking water pipes.

Paint

Lead is no longer added to paint except for a few specialist primer paints.

Old, flaking paint may be swallowed by children who pick it off with their fingernails and may swallow it if they put their fingers in their mouths.

Paint removal using blow lamps may produce lead vapours, which can be inhaled. Hot air paint removal guns use lower temperatures and are less likely to produce lead vapours.

Petrol

Tetra ethyl lead (TEL) used to be added to petrol as an anti-knock agent and valve lubricant. The lead would be present in exhaust fumes as a fine dust. This could be inhaled or swallowed.

Leaded petrol has been phased out in most countries as other anti-knock agents have replaced TEL.

Alternative fuels such as diesel, LPG, hydrogen and alcohol do not need lead additives.

Mercury

Mercury is found in three chemical forms: elemental liquid mercury, inorganic compounds and organic compounds. Their relative toxicities are due to the ease with which they are absorbed into the body. Once inside the body they are chemically changed, usually to Hg^{2+}.

- Liquid mercury is not efficiently absorbed in the gut, but it is easily absorbed in the lungs as a vapour.
- Inorganic mercury compounds are absorbed in the gut and through the skin. Not much inorganic mercury enters the brain as it is not liposoluble and cannot pass through the cell membranes.
- Organic mercury compounds are easily absorbed in the gut or in the lungs as vapours. Because they are liposoluble they can cross cell membranes and pass into the brain, across the placenta or into breast milk.

Mercury compounds in the environment may be converted into the more dangerous organic forms, e.g. by anaerobic bacteria in lake sediments, which convert inorganic mercury into methyl mercury.

▇ Noise pollution

Noise is unwanted sound that causes disturbance, annoyance or damage.

The effects of noise pollution

The effects of noise on living organisms

Excessively loud noises cause a range of problems to humans:

- deafness – especially to high frequency sounds
- stress, nervous disorders, heart disease, ulcers, irritability, headaches
- behavioural changes, e.g. aggression.

Quieter noises can also cause problems if they are annoying, e.g. a high pitch or rhythmic sound.

Loud noises may disturb farm livestock causing injury or breeding failure.

Nesting birds may abandon nests causing egg or chick losses and increased predation.

The effects of noise on non-living things

Sounds that cause an object to vibrate at its natural **resonant frequency** may eventually cause stress cracks to appear, resulting in structural damage. This is called **acoustic fatigue**.

Repetitive vibration can cause problems such as damage to buildings, bridges and underground pipes caused by the wheel vibration of trucks.

Sonic booms from supersonic aircraft can cause damage, e.g. to windows.

Sources and control of noise pollution

Industrial noise

Sources of industrial noise

Equipment involving impact between metal objects or the explosive expansion of gases pose a particular problem, e.g. stamping machines, air compressors.

Fig. 11.27 *Even remote, cliff-nesting birds' apparently safe environments can be vulnerable to noise pollution*

▇ Key terms

Resonant frequency: the frequency at which an object naturally vibrates.

Acoustic fatigue: stress cracking of a material caused by repetitive vibrations induced by sound.

Domestic noise pollution

Domestic noise pollution is mainly caused by domestic appliances such as washing machines, audio-visual equipment, garden equipment and barking dogs.

Although these may seem mundane, they can greatly reduce quality of life and are important issues.

Ways to control domestic noise include the following:

- keep audio equipment at a reasonable volume
- locate domestic appliances where they will cause least nuisance, e.g. the washing machine in a utility room rather than in the kitchen where people are more likely to be
- wear ear defenders when using power tools
- sit away from speakers at loud concerts
- volume limiters on disco equipment
- buy quieter appliances. This requires informed customers and the availability of the information to aid appropriate choices
- considerate control of pet dogs.

Measuring noise pollution

Loudness/volume

The deciBel scale

Sound intensity can be measured using a range of physical measures, but a convenient unit for measuring sound volume for hearing is the Bel. This is based on the sensitivity of human hearing so that the quietest sound that can be heard is 0 Bel. This level is called the 'threshold of hearing'.

To avoid having decimal values of Bels, the unit is divided into 10ths.

1 Bel = 10 decibels (dB)

The Bel is a logarithmic unit so a change of 1 Bel (10dB) represents a ten-fold increase in sound intensity. A 0.3 Bel (3 dB) change represents a two-fold change.

For specific applications the **deciBel** scale may be modified.

The dB(A) scale

The **dB(A)** scale is used for human hearing because it takes into account the sensitivity of human hearing by 'weighting' sounds between 1000 and 4000 Hz since the nerve cells in the ear are particularly sensitive to these and they are therefore heard most easily.

Measuring road traffic noise pollution

$L_{10}18_h$

The measure $\mathbf{L_{10}18_h}$ assumes that traffic noise is only a problem during the 18 hours between 6 a.m. and midnight. The noise level is measured during the noisiest 10 per cent of each of these 18 hours. The mean of these 18 levels is then calculated.

Traffic noise index (TNI)

The **traffic noise index (TNI)** is a measure that takes into account the noise levels exceeded for 10 per cent (noisiest) and 90 per cent (quietest) of the time.

Key terms

deciBel: the unit of sound used to monitor noise pollution.

Resonant frequency: the frequency at which an object naturally vibrates.

dB(A): the decibel scale, modified to weight the frequencies to which the human ear is most sensitive.

$L_{10}18_h$: measure of noise pollution used to monitor road traffic noise.

Traffic noise index (TNI): a measure of noise pollution used to monitor road traffic noise.

Measuring aircraft noise pollution

Noise and number index (NNI)

The **noise and number index (NNI)** measures the number of flights and the noise levels of aircraft above 80 dB as it was found that aircraft below 80 dB did not cause serious annoyance.

■ Ionising radiation

Some atomic nuclei are naturally unstable and undergo spontaneous reorganisation of the nucleus, which releases ionising radiation. Types of atoms that emit ionising radiation are called radionuclides. Ionising radiation can be produced artificially in X-ray machines and in nuclear reactors.

Ionising radiation is energy or particles emitted from a source, which, when they are absorbed, produce charged atoms or groups of atoms (ions), which would not otherwise have been there. Because they would not have been there, they may cause abnormal chemical reactions to occur or change the chemical properties of the material that absorbed it.

Radioactive decay and half-lives

All atoms of a particular unstable isotope have the same probability of decaying during a particular period of time. The likelihood of an atom decaying is not affected by the number of atoms present. So, for a mass of a particular isotope there is a predictable rate at which the atoms will spontaneously decay and emit ionising radiation. This rate is often expressed as the half-life, which is the length of time it takes for half the atoms present to decay. Note that 'half' relates to the total number of atoms at the beginning of that time period. So, if a mass is observed for several half-lives then the remaining proportion will be 1/2, 1/4, 1/8, 1/16, 1/32, 1/64, etc. of the original amount.

Different isotopes may have very different half-lives.

Table 11.4 *The half-lives of selected isotopes*

Isotope	Half-life	Source
Uranium 238	4.5×10^9 years	Naturally occurring in rocks
Iodine 131	8 days	Fission product of nuclear reactors
Caesium 137	30 years	Fission product of nuclear reactors
Plutonium 239	24,400 years	Produced in reactors by neutron bombardment of U238
Strontium 90	29 years	Fission product from nuclear weapons

Half-lives and health risks

Isotopes with short half-lives release all their radiation quickly and so are dangerous, but they do not pose a danger for long so short-term precautions may be sufficient to protect workers and the public.

Isotopes with long half-lives pose a danger for a long time but they do not emit much radiation in any given time period so the level of danger may be quite low.

Asexual reproduction in plants – vegetative propagation

Some plants, such as the strawberry, naturally reproduce asexually. This is known as **vegetative propagation.** Shoots grow out from the 'parent' plant. Where they touch the ground, roots and leaves start to grow and a 'daughter' plant develops.

Artificial asexual reproduction

There are two methods of artificial asexual reproduction: cuttings and micropropagation.

In many plant species, a new plant can be produced by taking cuttings; that is putting the cut end of a stem into damp soil. Roots develop from the cut end and a new plant is produced that is genetically identical to the original plant. The use of plant hormones to encourage root development increases the survival rate of cuttings.

Micropropagation involves growing plants from small pieces of plant tissue on a sterile nutrient medium under carefully controlled conditions. The medium is usually agar jelly containing sugars and mineral nutrients. Thousands of genetically identical plants can be produced much more quickly than by normal breeding programmes that produce great genetic variation.

Asexual reproduction in livestock – cloning

No commercially important livestock species normally reproduce asexually. So, to produce offspring that are genetically identical to an individual, artificial methods would have to be used.

Cloning using genetic material from body cells involves taking the genetic material from a body cell of one individual and implanting it into the egg cell from a different individual so that it develops like a normal fertilised egg.

Cloning techniques are still being developed. It is very unlikely that cloning will replace normal sexual reproduction of livestock, but it may be used for particular purposes. Some applications of livestock cloning may be the replacement of valuable animals that die, to increase the number of the best animals and to replace herds that have been culled during a disease epidemic.

Sexual reproduction

Sexual reproduction involves the combination of genetic material from two individuals to produce offspring that have characteristics similar to both parents but that are not entirely identical to either parent.

Sexual reproduction is essential for evolution as it produces individuals with new and unique combinations of genetic material. However, the offspring are often small, isolated and vulnerable, and may have a poor chance of survival.

Animal breeding

Selective breeding

Selective breeding may be used to exaggerate a single desirable characteristic or to combine several characteristics.

Crossbreeding between different breeds can be used to combine desirable characteristics in a hybrid breed.

Crossbreeding also reduces the problems of homozygous recessive genes found in inbred varieties. This is called '**hybrid vigour**' or **heterosis**.

Did you know?

The strawberry plant also reproduces sexually to produce fruit.

Key terms

Vegetative propagation: asexual reproduction.

Cloning: an artificial form of asexual reproduction.

Crossbreeding: producing offspring by mating parents of two different breeds or varieties.

Hybrid vigour: the good health achieved by breeding between breeds that are not closely related. This reduces the risk of inbreeding and recessive gene diseases.

Heterosis: the hybrid vigour produced by breeding between two organisms that are not closely related.

A number of methods can be used to manipulate breeding and increase its effectiveness or productivity.

Fig. 13.5 *Highland cattle can survive a harsh climate*

Artificial insemination (AI)

Artificial insemination (AI) allows the semen from one male to be used to impregnate many females. The method was originally developed for dairy cattle, but is also used for pigs and poultry.

The sperm is collected and frozen in an antifreeze solution to protect it. It can be used on site or transported to other locations. The frozen sperm can be stored for long periods.

'Extenders' are chemicals that make fertilisation more effective, so fewer sperms are needed and one semen sample can produce more offspring. By artificial insemination a bull can produce up to 10,000 calves per year.

Embryo transfer

Embryo transfer enables a female to be the mother of many more offspring than she could naturally produce.

A pedigree cow (with desirable characteristics) is given an injection of a hormone to stimulate ovulation and the release of eggs. These are washed out of her uterus and collected. They are then fertilised using semen from a pedigree bull (also with desirable characteristics). This type of fertilisation outside the female is known as *in vitro* fertilisation (IVF).

The embryos produced can be implanted into a number of fertile cows.

Each release of eggs can produce several offspring instead of one and there is no need to wait for pregnancy to end before more young are produced.

Plant breeding

Selective breeding

The selective breeding of plants has been carried out for centuries and has produced plants that bear little resemblance to their ancestors. Some plants that have changed or enhanced characteristics include maize with large, sweet seeds, disease-resistant potatoes and high protein-content potatoes.

Case study

The Green Revoluton

In the 1960s a lot of research was carried out to improve the qualities of cereal crops, especially rice. The new varieties such as rice variety **IR8** grew much faster, had short, stiff stalks that were less prone to wind damage and a greater proportion of the crop biomass was harvestable grain. Several harvests could be produced in one year, instead of a single one.

This greatly increased harvests and reduced famine problems, especially in India. However, farmers could only increase yields if they could afford the greater inputs of seeds, fertilisers, irrigation and pesticides.

Before the **Green Revolution** many peasant farmers paid low land rents as the landowners could not get higher rents in other ways and there was little demand for land.

After the Green Revolution much greater incomes could be achieved but only for those that could afford the required inputs. Poorer farmers did not benefit from increased incomes and could not afford the higher rents, so they lost their farms.

For a short time, the newly landless farmers gained employment on farms dealing with the increased harvests. But the increased farm incomes were often spent on machinery so the number of farm workers went down.

The landless unemployed ex-farmers and their families often moved to cities in the hope of gaining alternative employment. Where the Green Revolution was implemented with the allocation of land to the poor and financial support for investments there were fewer social problems.

Key terms

IR8: a Green Revolution rice variety.

Green Revolution: the agricultural changes since the mid 20th century, where high yielding cereal varieties were bred to increase food production.

Genetic engineering: the method of altering an organism's genetic makeup by artificially introducing genes from another organism, often of another species (see Genetic modification, page 225).

Transgenics: the process of artificially transferring genetic material from one organism into an individual of another species.

Gene-pool problems and inbreeding

If a very small gene pool or closely related population is used for breeding (inbreeding) then there is an increased probability of undesirable recessive genes producing offspring with undesirable characteristics.

This is a powerful argument for wildlife conservation to protect the wild relatives of crops and livestock that may be used to introduce new characteristics and maintain 'hybrid vigour'. They may also be the source of completely new desirable characteristics.

Genetic engineering/transgenics – GM crops

Genetic engineering/transgenics involves the transfer of genes for chosen desirable characteristics from the species where they are naturally found into a crop species to give it the desired characteristic.

One example is herbicide resistance as found in 'Roundup-ready' soya beans. The soya bean plants are resistant to the herbicide glyphosate (Monsanto trade name: 'Roundup'), which can be used to kill weeds without harming the crop.

Another example is insect pest resistance. The bacterium *Bacillus thuringiensis* produces a toxin that kills insects. The gene has been transferred to crops that can then produce the toxins against insect pests. For instance, BT corn produces toxins against the corn borer and BT cotton produces toxins against the cotton bollworm.

Golden rice

Vitamin A deficiency is a major nutritional problem in many tropical countries where rice is a staple food.

Golden rice includes genes from a daffodil and a soil bacterium, which produce beta-carotene. This forms vitamin A when the rice is eaten.

Advantages of GM crops

■ GM methods allow single characteristics to be introduced into a crop variety without the time needed for normal selective breeding techniques where it can take several generations to breed in the desired characteristics while eliminating undesirable characteristics that are also introduced.

■ GM methods allow desirable genes to be introduced from other species. This could never be achieved by selective breeding within the crop species itself.

■ If GM crops are pest-resistant then pesticide use may be unnecessary, which will reduce the environmental damage caused by the pesticide killing non-target species.

Disadvantages of GM crops

■ There are some claims that people eating GM foods suffer food-related allergies.

■ **'Horizontal gene transfer'** may occur where bacteria and viruses may combine the inserted gene into their own genetic material, thereby gaining new characteristics. Antibiotic-resistance genes, which have been introduced into crops, may be transferred to human pathogens and therefore produce 'superbugs' that would be more difficult to treat.

■ Pollen from GM crops may be transferred to wild plants of closely related species, which would then gain characteristics they would not naturally get. Weeds could gain herbicide-resistant characteristics.

■ The pollen from GM BT corn is blown onto the leaves of wild plants, which become toxic to the insects that eat the leaves, such as monarch butterflies.

■ Pollen from GM crops may be transferred to organic crops and this could invalidate their organic status.

■ The high yields that were initially produced have not been maintained.

Hormonal growth control

The growth and development of living organisms are controlled by hormones. By giving particular hormones, it may be possible to control growth and development.

Plant growth regulators

Auxins help vegetative propagation by stimulating root growth in cuttings. High doses cause uncontrolled growth and death so they can be used as herbicides such as 2,4 D (a synthetic auxin).

Gibberelins have several biological functions that can be used advantageously:

■ they inhibit shoot growth – used on cereals to produce shorter stems
■ they stimulate seed germination

- they inhibit root growth so more energy goes into fruit or seed growth
- they increase fruit size and grape cluster size.

Ethene or **ethylene** stimulates the ripening of fruit such as bananas.

Animal hormones

Animal hormones can be injected into livestock to increase the rate of some aspect of development. This may be economically beneficial, but there can be disadvantages.

Case study

BST

The hormone **BST** (bovine somatotropin) can be given to dairy cattle to increase milk yield. Its use is banned in many countries because of possible health effects on the cattle and uncertainty of its health effects on humans.

Anabolic steroid hormones

Anabolic steroid hormones are based on steroid sex hormones such as testosterone (male) and oestrogen (female). They increase the growth rate and gross growth efficiency of livestock.

Growth rate is a measure of the growth (increase in body mass) per unit time.

Gross growth efficiency (GGE) is a measure of the efficiency with which food eaten by livestock is converted into body mass. Anabolic steroid hormones can increase the GGE by up to 10 per cent. The use of steroid hormones in livestock has caused human health problems.

If residues of the steroid hormones remain in the meat when it is eaten then human growth may be affected. The synthetic steroid hormone stilbestrol was used in the USA in beef and poultry production. Stilbestrol is a synthetic derivative of the female hormone oestrogen. Some people who ate meat in which the hormone remained were affected. Girls experienced premature puberty and men developed female secondary sex characteristics.

Reduced competition – pest control

A pest is any organism that reduces the quality or productivity of crops or livestock in any way. Pests cause damage in the following ways:

- reduced harvest due to ill health of the crop or livestock
- reduced harvest due to competition for water, light, nutrients etc.
- kill the crop or livestock
- reduced quality of harvest – taste, appearance, storability
- reduced quantity of harvest – they eat it
- cause human disease.

Endemic and epidemic pests

Endemic pests are always present, usually in small or moderate numbers.

Epidemic pests are not normally present but there may be 'outbreaks' where they rapidly become a major problem.

Key terms

Ethylene: chemical that stimulates fruit ripening.

BST: bovine somatotropin – an animal hormone used to stimulate milk production.

Anabolic steroid hormones: a female or male hormone used to increase livestock gross growth efficiency.

Endemic pests: a pest that is normally present.

Epidemic pests: a pest that is not normally a problem but may become a serious pest when the population suddenly increases.

Hint

The term 'endemic' is also used in wildlife conservation, where it has a different meaning: an endemic species is only found in that area.

Table 13.2 *The main pest groups*

Pest group	Examples
Insects	Locusts, aphids (greenfly and blackfly) large white butterfly
Weeds	Wild oats, dock, ragwort
Fungi	Rust, smut, late blight
Molluscs	Snails and slugs
Nematode worms	Potato nematodes
Birds	Sparrows, pigeons, starlings, geese
Mammals	Rabbits, deer

■ Key terms

Monoculture: the growth of a single type of crop, usually over a large area.

Fig. 13.6 *Large fields with a single crop are more at risk of pest damage*

Indigenous and introduced pests

Indigenous species are native to the area where they are found. Many pests have been introduced from the area where they are naturally found. These are often more of a problem as they may not have any predators in the new areas.

The need for pest control

Roughly one-third of the world's food harvest is lost due to pests and diseases. Losses are particularly serious in LEDCs.

Pest control is especially important if intensive or large-scale farming methods are used; namely in monocultures and intensive rearing.

Monocultures

Growing a large area of a single crop (a **monoculture**) may result in a pest infestation being more serious as the pests can spread easily throughout the whole area. If the same area of crop were grown in smaller, dispersed areas then it would be less likely that the pests would colonise all the areas.

Intensive rearing

The intensive rearing of livestock involves keeping large numbers of animals close together. This makes it easier for pests and diseases to spread rapidly.

Chemical pest control using pesticides

Pesticides are toxic chemicals that are used to kill pests. They can provide very effective rapid control but there can be longer term problems caused by their use.

Pesticide application methods include:

■ spraying of crops in fields
■ powder coating of seeds
■ soil injection to kill pests in the soil
■ dipping of livestock to kill skin parasites
■ drenching where livestock drink the pesticides to kill internal parasites.

Pesticide properties

An understanding of the properties of a pesticide can help to predict the way it can be used to reduce pest damage and also how it may cause environmental damage.

Fig. 13.7 *The 'tramlines' in a field show where tractors travel, e.g. when they spray pesticides*

Specificity

Some pesticides are 'broad spectrum' and kill a wide range of species. Others are 'narrow spectrum' and kill a smaller range of species because they are more specific. Broad-spectrum pesticides are more likely to kill non-target species.

Persistence

Persistent pesticides break down slowly so they are more likely to remain in the environment long enough to travel long distances or travel along food chains.

Toxicity

Pesticides are intended to be poisonous to kill pests, but they may also be poisonous to other organisms. The difference in toxicities to the pests and to non-target species is important in assessing the likelihood of the pesticide causing environmental damage.

Cumulative pesticides/bioaccumulation/ biomagnification

Persistent, **liposoluble** pesticides can build up in organisms and therefore up food chains. The concentration may build up until it becomes toxic to non-target organisms.

Fig. 13.8 *Pesticide spraying*

Case study

Otters and pesticides

Otters in lowland Britain became very rare following the introduction of **organochlorine pesticides** such as dieldrin and aldrin. They are persistent, liposoluble substances that biomagnified along the food chain until the otters were poisoned. The pesticides were banned in the 1980s. Otter populations have gradually recovered as the remaining toxins have broken down.

Fig. 13.9 *The lowland British otter population is now recovering from pesticide poisoning*

Mode of action

Surface acting pesticides are sprayed on the crop and protect it by providing a toxic coat that kills pests if they land on the crop or eat it. However, unsprayed parts of the plant are not protected, nor are new leaves that grow after the crop was sprayed.

Systemic pesticides are absorbed then translocated throughout the plant. So, all the plant is protected until the pesticide degrades, but when the crop is eaten it is possible pesticide residues may still be present in the food. Washing the food will not remove it.

Pests of plants

Weeds

Weeds reduce crop yields by:

- competing for nutrients, water, light
- contaminating the harvested crop and spoiling the taste
- providing food for other pests, e.g. insects and fungi.

Weeds are often controlled by using chemical weed-killers or 'herbicides'.

Key terms

Liposoluble: the property of a substance dissolving in lipids.

Organochlorine pesticides: persistent insecticide group, e.g. DDT, dieldrin, aldrin. Most are now banned or restricted.

Systemic: a substance that is absorbed and transported throughout an organism.

Types of herbicides

There are two main types of herbicides: hormone herbicides and contact herbicides.

Hormone herbicides are systemic chemicals that kill weeds by modifying some aspect of growth or development in a way that is harmful to the plant, e.g. causing unrestrained growth until the plant runs out of food reserves and dies.

Particular hormones only affect certain plant types, so they can be used selectively, e.g. to kill broad-leaf weeds in cereal crops.

Examples of hormone herbicides include:

- 2,4 D – kills broad-leaf weeds by causing excessive cell elongation
- 2,4,5,T – this kills all broad-leaved plants but it has been used to control weeds in woodlands. Being insoluble in water, it was not washed down in the soil and only killed the shallow-rooted weeds while leaving the deeper-rooted trees unharmed. It is rarely used now as it is teratogenic to humans.

Contact herbicides 'scorch' the leaves on contact, damaging cell membranes and leaf surfaces. They can be used to control broad-leaf weeds in cereal crops as the weeds have leaves that are more easily wetted, while the herbicide runs off the cereal leaves, e.g. glyphosate.

Insects

Insects are an especially serious problem in the tropics where there are no cold winters to reduce populations.

Some insect pests are indigenous, but many have been accidentally imported from other countries. With no natural predators their populations may grow rapidly. Insects cause damage in the following ways:

- they reduce harvests by eating the crop
- they spoil the appearance of the crop
- they spread diseases such as sleeping sickness of cattle or virus diseases of crops.

Types of insecticide

Types of insecticides include organochlorines, organophosphates and pyrethroids.

Case study

Organochlorines: DDT, dieldrin, aldrin

Mode of action: DDT opens sodium gates in nerve cells – nerves keep firing – insect is stimulated to death.

Problems: organochlorines are very toxic to insects but have low toxicity to mammals. This was seen as a big advantage until it was discovered that other properties meant they could cause problems. Organochlorines are persistent, liposoluble pesticides. They are absorbed by living organisms and build up along food chains until animals such as birds of prey, otters and pelicans are poisoned.

common where all the trees in a large area were cut down. The lower value trees may be shredded and made into chipboard or sawn and used for plywood, forklift pallets or making moulds for concrete.

There is increasing pressure to use more sustainable exploitation techniques.

Tropical timber plantations

The deliberate growth of tropical trees may help to reduce shortages and reduce the unsustainable exploitation of natural forests. The plantations may be sustainable in that timber production may be carried on indefinitely but there may still be a lot of environmental damage.

Main features of tropical plantations

Habitat loss

The existing natural forest may be removed by clear-felling to provide the land for the plantation.

Fig. 15.6 *A teak plantation on an area of cleared rainforest in Central America. Teak are Asian tree species*

Monoculture production

Production is easiest to manage if a single species is grown. Growth rates will be similar so activities such as thinning and harvesting can be planned more easily. However, the low biodiversity and simple age structure of a monoculture produces a much lower wildlife value than the original forest.

Species selection

The species that are grown are high value species such as teak and mahogany or fast growing species like eucalyptus. They are often exotic non-indigenous species. This reduces the risk of pest attack, but also reduces their wildlife value.

Selective breeding

The trees that are grown are usually selectively bred for desirable characteristics such as fast growth, good appearance and pest resistance. The reduced gene pool of growing genetically uniform trees means they will have similar growth rates and physical properties, which will make processing the timber easier.

Sustainable forestry

Case study

Programme for Belize – a study of sustainable forestry

Programme for Belize (PfB) is a private, non-profit organisation that manages 100,000 hectares (250,000 acres) of rainforest in Belize in Central America called the Rio Bravo Conservation and Management Area. The main priority is the conservation of the rainforest and the sustainable use of its natural resources.

The establishment of the rainforest reserve is good for wildlife but could threaten the local communities that used to use the area for fuelwood supplies, timber for house building, farming clearings, hunting for food and the collection of medicines and fibres. A range of economic activities have been developed or encouraged. Local people are employed in forestry and ecotourism activities within the reserve. Ecotourist groups visit local restaurants, are fed on local produce and buy locally made souvenirs.

Key terms

Programme for Belize (PfB): independent conservation organisation that manages the Rio Bravo Conservation and Management Area in Belize, Central America.

Fig. 15.7 *The canopy of the rainforest*

Fig. 15.8 *Narrow access paths avoid important trees*

Fig. 15.9 *Sustainably logged mahogany trees*

Fig. 15.10 *Selective logging creates small clearings that are rapidly recolonised*

Some of the reserve is managed to investigate the effectiveness of the rainforest in carbon sequestration to counteract the releases of carbon dioxide from fossil fuel power stations.

The forests of Belize were selectively logged for mahogany by the British when it was the colony British Honduras. Although there are fewer mahogany trees than would be found in the natural forest, most of the forest is intact. Programme for Belize aims to sustainably log selected species while protecting the rest of the forest and increasing the number of mahogany trees.

Mahogany is the wood that is in most demand but PfB are trying to encourage the use of other species to increase trade without increasing the pressure on the mahogany population.

Trees are only selected for felling in areas with other trees of the same species with suitable seed trees upwind as mahogany has wind-dispersed seeds.

Access paths are as narrow as possible and avoid important trees, including their root systems.

The use of heavy machinery use is limited and does not include bulldozers. Each tree has a unique identification tag, which allows the timber to be tracked from its original location to its final retail destination such as a DIY superstore in the UK.

Surplus timber offcuts are supplied to local craftsmen, e.g. buttress roots are carved into tourist souvenirs or branches for furniture making.

A tree nursery is used to replant within the reserve and to provide saplings for local communities.

Reserve rangers patrol the reserve to control illegal logging.

Key terms

Forestry Stewardship Council (FSC): an organisation that certifies sustainable forestry operations.

The Forestry Stewardship Council (FSC)

The **Forestry Stewardship Council (FSC)** monitors timber production and accredits those producers that manage sustainable logging operations.

Deforestation

The loss of forests around the world should be of concern to everyone as they can produce resources and life-support system services for all of us, although it is the local population that will probably notice the loss most.

The causes of deforestation

Forests are renewable resources that can regrow, even if they are cut down. Deforestation will occur in any location where the rate of removal exceeds the rate of recovery and regrowth. This may involve a slow process of exploitation over many years that gradually removes the trees so that mature forest is degraded to young woodland, then scrub woodland and finally total clearance. Or it could involve the creation of increasing numbers of small clearings that gradually turn the unbroken forest into a patchwork of small forest areas that may finally disappear. In these cases it is hard to estimate the actual area of forest and is therefore difficult to estimate the rate of deforestation. The most dramatic deforestation involves large scale mechanical clearance using bulldozers, chainsaws and trucks to take out the logs. Such areas can be measured more easily.

The obvious solution to deforestation is simple: stop cutting trees down. The reality is more complex: we need to understand why the deforestation occurs and find ways of preventing it. A simple ban on felling may cause great hardship for those involved, often the rural poor in LEDCs.

Forests and maximum sustainable yield (MSY)

The trees in a forest will continue to grow until they reach maturity when the old branches that fall off and decay may balance new growth. When a tree dies and falls over it will create a clearing in which young trees, that might not otherwise have survived, get the chance to grow and fill the gap.

Taking out trees or groups of trees may be sustainable as the gaps are recolonised in the same way as a natural clearing. Larger clearings may require active replanting as natural recolonisation may be much slower.

Deforestation will occur if the rate of clearance is faster than the maximum rate of regrowth (natural or aided by humans). The maximum sustainable yield (MSY) can be increased by using an agricultural approach to forest production:

- select species that grow quickly
- use selective breeding to enhance growth and quality
- control tree spacing
- increase nutrient supplies
- control pests and predators.

This may increase the MSY, but not necessarily the broader environmental sustainability.

In many cases forests are cleared with no intention of replanting or allowing it to regrow.

Timber

Commercial logging

The commercial logging industry produces timber for industrial processing and possibly for export.

> **Hint**
>
> Revisit what you learnt in Unit 1 about ecological succession to understand how small clearings can benefit wildlife but large ones can cause damage.

Wood from boreal taiga forests in Canada and Scandinavia is processed for building construction and paper and furniture manufacture. The controls on felling and forest management ensure sustainability.

Exports of timber from tropical forests are rarely sustainable. Felling may be large scale with no replanting.

Subsistence wood use

Subsistence wood use refers to where wood is used by the local community to build houses, furniture or as a fuel. In LEDCs the people may not be able to afford alternative materials or fuels. The solution here may be to encourage sustainable logging with replanting.

Farmland

Cash crop commercial agriculture

Fig. 15.11 *Rainforest clearance to create farmland*

Growing world populations, rising affluence and a better transport infrastructure can make almost any large area of land with reasonable soils and climate valuable for food production. The profits can be great, especially if land and labour costs are low, as is the case in LEDCs.

In tropical areas the soils may be too poor to sustain arable agriculture for more than a few years.

The growth in demand for beef, arable crops and biofuels such as palm oil increase the demand for farmland. Even if the producers do not clear forest themselves, they are contributing to the land shortage that encourages others to do so.

Subsistence farming

Where poor people in LEDCs are growing food to feed their own families, restricting access to land may increase poverty. The solution here may be to help maximise production on the land that is used rather than clear more and help with the social improvements that raise living standards and slow population growth.

Many of the families that clear rainforest to farm used to live on the land that was taken over by commercial agriculture.

Case study

Brazil in the 1980s

The rapid expansion of commercial agriculture in the 1980s helped to create much-needed income, which helped Brazil deal with its huge international debt. But the peasant farmers lost their land and, with mechanisation, their jobs too. Many moved to the major cities in search of work but, with a global economic recession, jobs were hard to find. Social and political unrest in the slums led the government to encourage people to migrate to remote states such as Rondonia. They were given rainforest plots, which they cleared to create farm smallholdings.

While rainforest is still being cleared the Brazilian government now actively encourages conservation in many areas.

Heat exchanger

A heat exchanger allows the house to be well ventilated without losing too much heat.

Heating system

Good building design will minimise the need for additional heating, but, where necessary, it should use renewable sources such as solar power and heat pumps. If fossil fuels are chosen then high-efficiency appliances should be used.

Orientation and passive solar gains

Orientation involves arranging the building so that the rooms that require the highest temperatures face the direction that will result in the greatest amount of solar heating.

Windows

Windows that face the sunny side of a building should be larger to maximise solar gains. Double or triple glazing will reduce heat losses.

Sunpipes

These are cylindrical pipes with highly reflective inner surfaces that can carry sunlight from a roof surface down to rooms that get little direct sunlight.

Appliances

Domestic appliances should be selected for their low energy and water use and long lifespan.

Occupancy sensors

Occupancy sensors can be used to turn off lights and other appliances when rooms are unoccupied.

Garden

Gardens can be designed and managed to benefit wildlife and compensate for the loss of habitat caused by the construction of houses.

Drainage

Impermeable surfaces such as patios, driveways and roofs reduce the infiltration of water into the ground. If this water is collected by a drainage system then it can increase flow rate surges downstream and increase flooding risk. This is a particular problem in large urban areas where the amount of water collected after heavy rain may be very large.

Heat storage

A comfortable home will have a reasonably constant, warm temperature but the availability of the heat source may not be constant. Some design features can help to minimise temperature fluctuations and reduce the need for extra heating during cold weather.

Construction with materials that have a high thermal mass, such as stone, will reduce temperature fluctuations. If the building is partially or completely buried or 'earth sheltered', then heat losses to the atmosphere are reduced.

A solar water heat store can use solar energy captured in the summer for use during the winter for space heating.

Activity

Many improvements in building design that would reduce energy use are expensive. Try to identify the improvements that would become viable if energy was much more expensive.

Fig. 16.14 *This impermeable driveway increases runoff when it rains heavily*

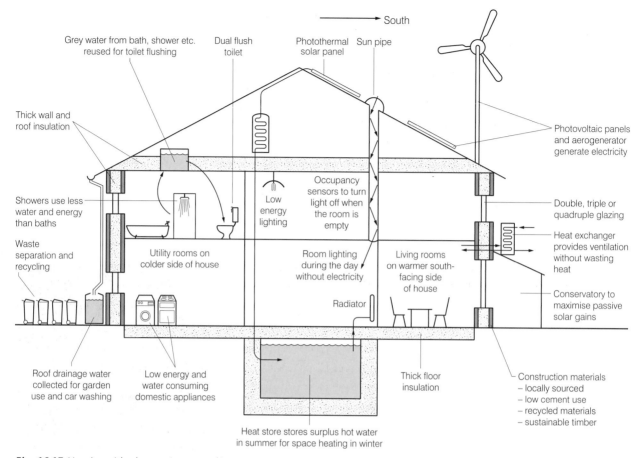

Fig. 16.15 *Housing with a low environmental impact*

Ecotowns

The construction of new ecotowns in the UK has been proposed as a way of tackling housing shortages with developments that have a lower environmental impact than more traditional housing schemes. Although many could be built on brownfield sites such as old military airfields or quarries, some greenfield sites would have to be used.

By being separate from existing towns ecotowns would need their own infrastructures, which could incorporate low-impact technologies. If new housing were built as an extension to an existing town then it is more likely that the old infrastructure would be used.

Water supply

A dual water supply could be used where the water would be treated to different quality standards depending upon its intended use. Water for low quality uses would require fewer purification processes, which would reduce the use of energy and chemicals that could cause pollution in their production or use.

Waste water

A dual waste collection system would use the full range of treatment processes for heavily contaminated water, such as toilet waste. Water from baths, showers and washing machines would require less treatment.

A combined heat and power district heating system

Burning fuel in a power station to generate electricity produces a lot of heat that is often wasted. In addition to providing electricity, the use of a combined heat and power station would provide hot water that could be distributed for the space heating of buildings. This would reduce the use of extra fuel for space heating.

Wildlife conservation in gardens and green spaces

Urban green spaces can be very valuable for wildlife, with a wide variety of habitats. If housing developments have a low housing density there may be more space for trees, ponds and other habitats that are missing from overcrowded developments.

There is often a general assumption that housing developments on greenfield sites should take the minimum possible land area. However, if (and only if) a larger area is used for low-density housing, there may be a greater habitat diversity with public green spaces so its wildlife value would be increased.

Transport systems

In a new development that is being developed from scratch, it is possible to include a comprehensive public transport system to reduce the inefficient use of personal transport such as cars. Car usage can be discouraged using methods such as limiting parking facilities and reducing speed limits, while public transport is made more convenient and reliable.

Streets in housing areas could have traffic calming measures, cycle paths and pedestrian priority to increase pedestrian safety and encourage travel without using a car.

Waste disposal and recycling

A new waste collection system could include the best features of other systems:

- separate bins for kerbside collection of glass, paper, card, metals etc.
- local collection points for other wastes, including separation of rubble, soil and materials in small quantities that would not justify kerbside collection such as engine oil, batteries and fluorescent lights
- community groups may be supported for the refurbishing of furniture and domestic appliances, or to provide environmental advice.

Fig. 16.16 *Traffic calming makes roads safer for cyclists and pedestrians*

The example of Kenya

Kenya is often seen as a success story of development compared with many other countries in Africa. However, it still faces many challenges, as a rapidly growing population, environmental degradation and lack of resources make progress difficult.

Wildlife conservation

The wildlife of Kenya is a major source of tourism income that can help the national economy, although the environmental cost of so much long-distance travel is high. The income generated by tourism justifies continued conservation of the wildlife and the habitats where it is found.

Ecotourism

The sensitivity of the main wildlife sites has led to the development of exclusive tourism with a small number of visitors paying high prices for the experience. The smaller number of visitors may reduce impact in some ways but the impact of luxury accommodation may be relatively high with high water and energy use for air conditioning, Jacuzzis and exclusive transport such as light aircraft.

Fig. 16.17 *Wildlife spectacles attract tourists and benefit the economy*

Fig. 16.18 *Large numbers of tourists can disturb wildlife*

Fig. 16.19 *Safari rally events generate income but can disturb wildlife*

Case study

Masai Mara Game Reserve

Unlike many national parks, the Masai Mara Game Reserve in Kenya is owned by the local council, so some of the income from tourism directly benefits the local community. However, the income does not fully compensate for the crop damage caused by game animals that move out of the reserve looking for food. It is illegal for the farmers to kill the game animals but fencing farmland restricts their movement and prevents access to feeding areas they need when there is not enough food within the park.

The income from rally car racing through game reserves produces valuable income but it is also a threat to the wildlife.

Coral reefs

Over half of the population of coastal areas of Kenya rely on fishing or tourism for their livelihoods but the coastal resources are being overexploited. Fish populations are overfished. The coral reefs are threatened by sediments from rivers, agricultural chemicals, and physical damage from fishing and tourists. Coral bleaching caused by high water temperatures is also a serious problem, especially during El Niño years.

The establishment of protected areas where harmful activities are restricted has produced some success at maintaining or increasing local biodiversity.

Agriculture

There is not much good farmland in Kenya and much of the best land is used to grow cash crops for export. This forces the growing rural population to farm land more intensively and therefore less sustainably and to cultivate marginal land that is not really suitable for farming.

The biggest problem is declining fertility and soil erosion, which are increasing as forests are cleared and marginal land is cultivated.

Fig. 16.20 *A tea plantation*

Case study

Lake Naivasha

Lake Naivasha is a large freshwater lake, which is important for wildlife such as hippos and over 400 bird species. Fisheries also provide food for the local people.

Introduced species such as the predatory fish, the Nile perch and the herbivorous North American red swamp crayfish have had serious ecological effects.

The biggest threat is the farming of flowers on the surrounding land, which are exported to Europe. Unsustainable irrigation is lowering the water level and runoff carrying fertilisers and pesticides is causing pollution.

Case study

Independent and dependent variables

The independent variable is the factor that is being deliberately altered and measured to see if it controls the dependent variable, which is also measured.

Other variables may affect the results and should be controlled if possible. If not, it may be necessary to measure them to investigate whether they are affecting the study.

Case study

Hypotheses and null hypotheses

The hypothesis states that a relationship exists, while the **null hypothesis** states that there is no relationship.

For example:

Hypothesis – burning more fuel makes it hotter

Null hypothesis – there is no relationship between the amount of fuel burnt and temperature.

It is important in scientific research to be cautious about conclusions. Rather than concluding that the results support the hypothesis, it sets a higher standard of evidence to conclude that it would be implausible to accept any other explanations.

Representative data

Careful planning is needed to ensure that the data collected are reliable in investigating whether a hypothesis should be accepted or rejected because results are not necessarily representative of the situation being investigated.

Many scientific investigations pose questions such as: 'What is the temperature/pH/salinity/turbidity etc. of ....?'

Such questions are often phrased as if there is a single answer, but the value collected from a single observation in a single location or at a single time may not be representative of the typical or long-term situation. Collecting data everywhere all the time would clearly give results that represented the true situation, but it is impossible to do this.

Some questions may have a single answer but it would be unrealistic to actually find it, e.g. 'How many cod are there in the North Sea?' or 'How much carbon is stored in the Amazon rainforest?'

In all of these situations, sub-samples must be taken, which can then be used to estimate the total value.

Sample location

If data are not being collected at a single location, then decisions will have to be made about the positioning of the sample sites.

It is essential to avoid introducing bias by deliberately selecting locations to support or dismiss a hypothesis. This can usually be achieved by random sampling, which can be done in several ways.

Hint

You will come across many statements that are used to discuss environmental issues, such as:

- For African elephants, the average lifespan in captivity was only 19 years compared with 56 years in the wild.

- Global average temperatures are predicted to rise by between 1.4 °C and 5.8 °C by 2100.

- Spring is coming about two weeks earlier than it would have been 30 to 50 years ago and autumn about a week later.

- Figures released by the (Brazilian) environmental ministry showed that between 2000 and 2001 the rate of destruction (of rainforest) fell by 13 per cent.

- Ice more than two years old now makes up about 30 per cent of all the ice in the Arctic, down from 60 per cent two decades ago.

(All statements from **http://news.bbc.co.uk** December 2008)

You should ask Who? How? Where? about the background to these statements. Even if a statement is valid, the margin of error of the figures should be considered.

Key terms

Independent variable: the factor that is deliberately altered or measured to see if it affects the dependent factor.

Dependent variable: the factor that may be controlled by the independent variable, i.e. the 'results' that are measured.

Null hypothesis: the no-link theory against which the hypothesis is being tested.

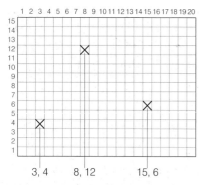

3, 4   8, 12   15, 6

Each sample is located using coordinates selected using two random numbers

Fig. 17.1 *Sampling in an area where a coordinate grid can be laid out*

Sample sites are selected using random numbers in the range 1–37

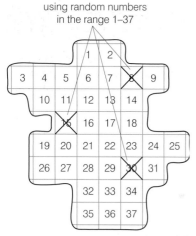

A study area where a simple coordinate grid cannot be used. Each possible sample site is given a number

Fig. 17.3 *A sampling area with random sampling sites*

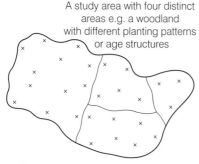

A study area with four distinct areas e.g. a woodland with different planting patterns or age structures

The total study area is divided into its distinct areas: 'strata'. The number of samples taken from each area is proportional to its area

Fig. 17.4 *An area that has been sub-divided for stratified sampling*

■ Random sampling – the sampling sites within the study area can be chosen using a table of random numbers or the random number function of a scientific calculator to select coordinates on a grid.

■ Systematic sampling – this involves samples taken using a chosen pattern or spacing. It can be considered to be random in that the samples are not chosen based on observable differences in the study area.

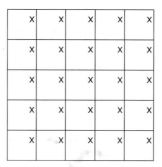

Systematic sampling in a study area

Sample locations are established at fixed distances and directions from the first sample location or in a fixed pattern.

The first sample may be located using random numbers to select coordinates.

Systematic sampling along transects

Belt transect

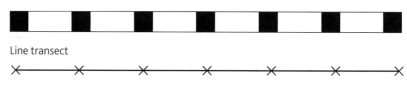

Line transect

The intervals (distance or spacing) between sample sites should be small enough that the results are representative without collecting more than required. This can be determined by carrying out a preliminary study.

Fig. 17.2 *Systematic sampling*

■ Stratified sampling – if a study area has clear sub-areas with differences that will influence the results then it may be necessary to study each area individually, then combine the results.

Number of samples

An individual sample may not be representative of the typical value that is sought: it may be anomalous, but it will only be possible to spot the anomalies if many results are collected and compared. Values may also fluctuate so a single value could be misleading. The ideal number of samples is that where collecting more would not significantly change the mean value of the results collected.

The number of results collected should also be planned with consideration for any statistical tests that will be carried out, as this will affect whether the tests can be used, and the degree of significance of the conclusions that can be made.

Sample size

Each sample must be large enough to be representative and minimise the variability of situations that are not **homogenous**.

Standardised sampling methods

The method that is used to collect data must be the same on each occasion to avoid introducing extra variables.

Sampling aquatic animals

Kick sampling

The riverbed is disturbed by kicking so that mobile invertebrates are washed into a net downstream.

Limitations of kick sampling are:

- disturbance by kicking cannot be standardised
- organisms attached to rocks will not be collected
- organisms that can swim may escape.

Surber samplers

Surber samplers develop kick sampling to produce more quantitative results. The net frame covers a fixed area of riverbed and the net sides reduce the ease with which mobile animals can escape. The riverbed is disturbed using a trowel or similar tool and stones are inspected manually.

Fig. 17.9 *A surber sampler*

Fig. 17.10 *Kick sampling*

Planktonic organisms

Phytoplankton and zooplankton that drift and are carried by currents can be sampled with plankton nets. Different mesh sizes are used for different sizes of organisms.

Sampling terrestrial and aerial animals

Pitfall traps

Mobile invertebrates that move over the ground surface can be trapped as they fall into the collection chamber. A cover is used to prevent rain, litter or larger animals entering.

Limitations of pitfall traps are:

- organisms beneath the surface or that hardly move will not be collected
- if a killing fluid is used, some species may be attracted while others are repelled
- if no killing fluid is used, predators may eat the other animals that fall in
- very active animals are more likely to be trapped.

> ### Key terms
>
> **Surber sampler:** an aquatic invertebrate sampling frame and net that provides more quantitative data than kick sampling.
>
> **Pooter:** a mouth-suction device to pick up invertebrates in soil or leaf litter.

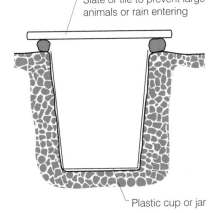

Slate or tile to prevent large animals or rain entering

Plastic cup or jar

Fig. 17.11 *A pitfall trap*

Fig. 17.12 *A pooter being used to collect invertebrates in a leaf litter sample*

Suction samplers

These allow the collection of invertebrates by using a lowered air pressure to pull them into a sample chamber or net. Examples include the following:

- **Pooter** – a small mouth-operated collector used for small visible invertebrates in soil or litter samples.
- Backpack suction samplers – these are larger motorised samplers that are used by professional ecologists to collect invertebrates in grasslands or similar habitats. If the tube diameter, air-flow rate and collection time are standardised, then quantitative data can be collected.

Fig. 17.13 *A light trap*

Fig. 17.14 *Tullgren funnels, with the lights removed*

Sweep nets

These are lightweight nets with a large diameter that can be swept through vegetation to collect invertebrates. If the number and length of the sweeps are standardised then semi-quantitative data can be collected.

Beating trays

A cloth sheet or plastic tray is placed beneath the bush or tree to be sampled. The branches above the tray are shaken vigorously or hit with a heavy stick. Invertebrates will fall onto the tray where they can be collected. Semi-quantitative data can be produced if the tray has a fixed size and a standardised beating technique is used.

Light traps

Some flying insects such as moths are attracted to lights and can therefore be trapped. Mercury vapour lamps are most effective but fluorescent tubes are easier to look at and do not get hot.

Tullgren funnels

Tullgren funnels allow the extraction of invertebrates from soil or litter samples. Invertebrates that move away from light or raised temperatures move downwards until they fall into the collection chamber.

The limitations of Tullgren funnels are:

■ immobile organisms are not collected
■ larger organisms may not pass through the mesh.

Collection of earthworms

Collection of earthworms by digging them up is very time-consuming. Flooding the soil with water containing an irritant chemical such as methanol or detergent forces the worms to come to the surface.

Limitations of using irritant solutions to collect earthworms are:

■ worms in deep soil may not be affected
■ some worms may escape sideways
■ different species or sizes of worms may react to the chemicals differently.

Photographic surveys

Where individual animals are seen, it may be difficult to be sure whether the population is large or whether a few individuals are being seen on many occasions. Some species have individuals with recognisable markings that can help identifications such as the tailfins of humpback whales, stripes of tigers or the ragged ears of elephants.

Tags

Studies of bird populations often involve catching live birds, which then have small metal rings placed on their legs so they can be identified if they are caught again. Bright numbered and coloured plastic wing tags can be used to identify individuals without having to catch them again.

Tagging is also used to study the movements of sharks, turtles, lobsters and whales.

Radio tracking allows the monitoring of movement, which helps to establish migration routes, feeding areas and territories.

Factors affecting noise levels

Noise levels can be measured with an electronic sound level meter but, if one is not available, many mobile phones incorporate sound level meters. If uncalibrated meters are used then the same one should be used again to collect all the results that will be compared with each other, or the degree of variation established by measuring a range of sounds with all meters before the actual results are collected.

Distance from source

The fact that noise levels decline as the distance from the source increases is predictable, but collecting quantitative data and assessing variations in noise levels from sources such as roads and airports are more difficult.

Because the decibel scale is logarithmic a simple arithmetic mean does not give a true representation of the typical sound levels.

The presence of sound absorbing or reflecting surfaces and topography can affect the noise levels, so the study sites should be as similar as possible.

Acoustic insulation

This may be investigated under laboratory conditions with a source of sound, such as a speaker, surrounded by acoustic insulation of different materials or thickness.

The effectiveness of insulation around roads or airports such as baffle mounds, embankments, walls, fences or vegetation can be compared.

Skills for ENVS4

The effect of slope and vegetation on rain splash erosion

Scientific experimentation relies upon good experimental design, which requires a knowledge of all the variables that could influence the results. Ideally, all the variables that are not being investigated should be standardised so that the only factor that affects the dependent variable (the results) is the independent variable that is being deliberately controlled. This is easier to do under laboratory conditions than under field conditions. Factors that cannot be standardised should be measured so that their influence on the results can be assessed.

> **Hint**
>
> There is no single method for assessing the effect of slope or vegetation on soil erosion. You would be assessed on your knowledge of the variables involved and how they could influence the results.

Variables that can affect rain splash erosion:

- rainfall intensity
- soil texture
- soil compaction
- organic matter content
- soil depth
- permeability of the material beneath the soil
- precipitation rate – volume per unit area per unit of time
- raindrop size
- raindrop height drop
- gradient
- vegetation cover
- root binding.

Many experimental procedures have limitations that are an unavoidable part of the method. It is important that these should be known and their impact on the validity of the results appreciated.

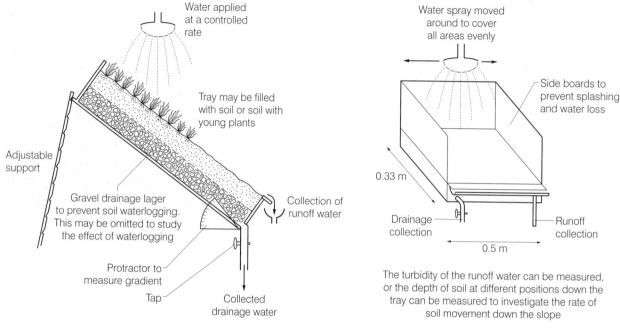

Fig. 17.21 *An experimental system for investigating rain splash erosion*

The effect of trees on microclimates

Trees affect the local climate in and around woodlands. Even individual trees produce microclimates. An understanding of how these microclimates are produced and the effects they have can alter management plans for urban areas and rural areas such as farmland, forestry plantations and woodlands for wildlife conservation.

Variables that should be considered

Vegetation features:

■ tree height
■ foliage characteristics
■ tree spacing
■ total area of woodland.

Features of the microclimate that can be measured:

■ wind direction
■ wind velocity
■ light levels
■ soil moisture
■ temperature.

To measure the effect of trees on microclimates the general weather conditions must also be known to identify local differences.

Summary questions

1 Outline the general principles of standardising the following in environmental studies:
   - sample location
   - number of samples
   - sample size
   - timing of sampling. *(4 marks each)*

2 Describe how the use of preliminary studies can help to increase the validity of results while preventing the collection of unnecessary data. *(10 marks)*

3 Outline the methods that can be used to monitor populations of terrestrial/aquatic populations. *(essay 20 marks)*

4 Describe the methods that can be used to compare features of two soils. *(essay 20 marks)*

5 Outline the methods that can be used to assess the effect of trees on microclimates. *(essay 20 marks)*

6 Describe how climatic variability could be monitored to assess the viability of solar power/wind power. *(essay 20 marks)*

Website reference list

General environmental issues

http://www.greenfacts.org/en/global-biodiversity-outlook/toolboxes/figure-2-18.htm

WWF

http://www.wwf.org.uk/

Friends of the Earth

http://www.foe.co.uk/

Greenpeace

http://www.greenpeace.org.uk/

Environment Agency

http://www.environment-agency.gov.uk/

NERC

http://planetearth.nerc.ac.uk/

UK environmental statistics: DEFRA 'Environment in your pocket'

http://www.defra.gov.uk/environment/statistics/eiyp/pdf/eiyp2008.pdf

The living environment (ENVS1)

Wildlife conservation

Wildlife Extra magazine - general conservation issues

http://www.wildlifeextra.com/index.html

Why wildlife conservation is important

http://www.countrysideinfo.co.uk/biodvy.htm

Zoological Society of London - field Conservation

http://www.zsl.org/field-conservation/

World Conservation Union - IUCN

http://www.iucn.org/

Red lists of endangered species

http://www.iucnredlist.org/

UNEP World Conservation Monitoring Centre

http://www.unep-wcmc.org/

Conservation methods

Legal protection and sustainable management
CITES

http://www.cites.org/

IWC

http://www.iwcoffice.org/

Captive breeding and release

Good general info + species case studies

http://nationalzoo.si.edu/ConservationAndScience/EndangeredSpecies/CapBreedPops/default.cfm

World Association of Zoos and Aquariums

http://www.waza.org/conservation/projects/index.php

Seedbanks

http://www.kew.org/msbp/

Conservation in the UK

General conservation in the UK - many links

http://www.naturenet.net/index.php

National Trust

http://www.nationaltrust.org.uk/main/

EU Birds Directive

http://www.birdlife.org/action/awareness/eu_birds_directive/what.html

Bird Life International

http://www.birdlife.org/

UK Biodiversity Action Plan

http://www.ukbap.org.uk/default.aspx

Natural England

http://www.naturalengland.org.uk/

RSPB

http://www.rspb.org.uk/

Woodland Trust

http://www.woodland-trust.org.uk/

Wildlife Trusts

http://www.wildlifetrusts.org/

Wildfowl & Wetlands Trust

http://www.wwt.org.uk/

Plantlife

http://www.plantlife.org.uk/index.html

Game & Wildlife Conservation Trust

http://www.gct.org.uk

Species, habitats, designated areas

http://www.jncc.gov.uk/

AONBs

www.aonb.org.uk

Interactive UK maps

http://www.bgs.ac.uk/mineralsuk/digital_maps/maps/home.html

Ramsar sites

http://www.ramsar.org/

Conservation abroad

Biomes

http://www.blueplanetbiomes.org

Tropical Rainforest

http://www.mongabay.com/

Coral reefs

http://oceanworld.tamu.edu/students/coral/coral5.htm

http://www.coralreef.noaa.gov/

http://www.mcsuk.org/

http://www.epa.gov/OWOW/oceans/coral

Great Barrier Reef

http://www.gbrmpa.gov.au/

Antarctica

http://www.antarctica.ac.uk/

http://www.coolantarctica.com/Antarctica%20fact%20file/science/human_impact_on_antarctica.htm

Land use conflicts in a National Park

http://www.peakdistrict-nationalpark.info/studyArea/factsheets/03.html

Green belts

http://www.naturenet.net/status/greenbelt.html

The physical environment (ENVS2)

The atmosphere

Development of the atmosphere

http://www.ux1.eiu.edu/~cfjps/1400/atmos_origin.html

Global climate change

http://www.ecobridge.org/content/g_tht.htm

Gulf Stream, North Atlantic Conveyor

http://www.noc.soton.ac.uk/rapid/sis/atlantic_conveyor.php

Ocean currents

http://science.nasa.gov/headlines/y2004/05mar_arctic.htm